34th PIANC World Congress 2018

Connecting Maritime Hubs Globally

Environment

Panama City, Panama
7 - 11 May 2018

ISBN: 978-1-5108-8703-9

Printed from e-media with permission by:

Curran Associates, Inc.
57 Morehouse Lane
Red Hook, NY 12571

Some format issues inherent in the e-media version may also appear in this print version.

Printed by Curran Associates, Inc. (2019)

For permission requests, please contact
The World Association for Waterborne Transport Infrastructure (PIANC)
at the address below.

The World Association for Waterborne Transport Infrastructure (PIANC)
General Secretariat
Boulevard Roi Albert II, 20 – B. 3
B-1000 Brussels
Belgium

Phone: +32 2 553 71 61

info@pianc.org

Additional copies of this publication are available from:

Curran Associates, Inc.
57 Morehouse Lane
Red Hook, NY 12571 USA
Phone: 845-758-0400
Fax: 845-758-2633
Email: curran@proceedings.com
Web: www.proceedings.com

TABLE OF CONTENTS

COMMITTEES

INTERNATIONAL ORGANIZING COMMITTEE (IOC)

LOCAL ORGANIZING COMMITTEE (LOC)

Mr. Manuel Benítez, Congress Co-Chair (ACP)
Mr. Rogelio Gordón, Congress Co-Chair (ACP)
Mr. Oscar Vallarino, Local Executive Committee (ACP)
Ms. María Teresa Maure (ACP)
Mr. Ricardo Álvarez (ACP)

INTERNATIONAL SCIENTIFIC COMMITTEE (ISC)

Dr. Philippe Rigo, Vice President of the SC (Belgium)
Dr. Michael Fastenbauer, Chair (Austria)
Mr. Jean Michel Hiver (Belgium)
Mr. Jose Sanchez (USA)
Mr. Wu Peng, Chair (China)
Mr. Jim Stirling, Chair (UK)
Mr. Fabrice Daly, Chair (France)
Mr. Rik Thomas (Belgium)
Mr. Juergen Troegl, Chair (Austria)
Mr. Leonel Temer (Argentina)
Ms. Milou Wolters, Chair (Netherlands)
Mr. Stefan Devocht (Belgium)
Dr. Carsten Thorenz, Chair (Germany)
Dr. Tom De Mulder, Chair (Belgium)
Mr. Holger Schüttrumpf (Germany)
Mr. John Clarkson, Chair (USA)
Mr. Timothy Paulus, Chair (USA)
Dr. Jasna Muskatirovic (Serbia)
Mr. Ian White, Chair (UK)
Mr. Francisco Esteban Lefler, Chair (Spain)
Mr. Ignacio Sanchidrian Vidal, Chair (Spain)
Mr. Hartmut Bruehl, Chair (Germany)

Mr. H.N. Aswath (India)
Mr. Craig Philip, Chair (USA)
Mr. Edward Schmeltz (USA)
Mr. Hans Burcharth (Denmark)
Dr. Mohammadreza Allahyar (Iran)
Ms. Katrina Dodd (Australia)
Mr. Esteban Biondi, Chair (USA)
Mr. Pablo Arecco, Chair (Argentina)
Mr. Nicholas Pansic, Chair (USA)
Dr. Todd Bridges (USA)

LOCAL SCIENTIFIC COMMITTEE (LSC)

Dr. Dimas E. Portillo L., President of the SC (ACP)
Mr. Abdiel Perez (ACP)
Mr. Rigoberto Delgado (ACP)
Mr. Antonio Abrego (ACP)
Mr. Ricardo Álvarez (ACP)
Dr. Ricardo Ungo (ACP)
Mr. Juan Wong, Chair (Panama)

COPYRIGHT AND REPRODUCTION

PREFACE

This book of abstracts contains 284 abstracts presented at the 34th PIANC World Congress held in Panama, Republic of Panama, from 7 to 11 May 2018. PIANC World Congress presents and discusses topics relevant to the waterborne transport and infrastructure sector. It rotates among PIANC's member countries every four years and it is open to members and non-members. Participants, have the opportunity to exchange knowledge and experiences.

PIANC's goal is to advance the sustainable development of shallow and deep-draft navigation issues including dredging and dredged material disposal, navigation and pot infrastructure, recreational navigation and related environmental matters.

The Book of Abstract is arranged according to the main Themes of the Congress namely: Inland Navigation, Dredging, Logistics & Infrastructure, Ports, Marinas, and Environment. Inside each Theme, the abstracts are arranged by the topics of the Theme.

Inland Navigation

- Waterway infrastructures: locks, weirs, riverbanks, others.
- Advances in navigation locks design after the Panama Canal Expansion
- Inland navigation channels: safety and reliability
- Integration of inland waterways into inter-modal supply chain
- Inland navigation, waterways, ports & terminals
- Systems and Infrastructures: design & management for inland navigation projects
- River Information Services (RIS, AIS, others)
- River navigation flow control

- Salt water intrusion
- Maintenance and operation of IW, transport & infrastructures
- Inland navigation in South America

Dredging (in the framework of port and navigation projects)

- Effective planning and execution of dredging projects
- Lessons learned from dredging projects worldwide
- Risk management in dredging projects
- Disposal site planning taking into account alternative land uses
- Current dredging & management innovations

Logistics & Infrastructure

- Challenges for capitalizing navigational channels bordering lands
- Synergies among airports, ports and inter-modal assets
- Integrated management of global supply chains
- Integrating ports and economic special Zones
- Provision of added value logistic services in maritime hubs
- Project management in the Panama Canal Expansion

Ports

- Impact and innovative solutions for hinterland ports
- Dry ports: better and more efficient
- Ports of the future: technologies, automation, traceability
- Port management models
- Maritime Port planning and operations
- Inter-modal connections
- Coastal and Port Engineering (in relation with navigation)
- Storm Surge & Tsunami barriers and Flood protection gates

Marinas

- Sustainable and resilient marina design-Marinas Working with Nature (best practices and case studies)
- Marinas as part of Tourism, Ports and Urban Master Plans (best practices and case studies)
- Planning Framework for Recreational & Tourism Navigation Infrastructure Systems (goals, strategies and challenges)
- Regulatory Framework and Private Marina Development in Countries in Transition (best practices and case studies)
- Regional Cooperation and Standardization of Recreational Navigation Information Systems Transition (best practices and case studies)
- Recreational components of mitigation plans for large navigation infrastructure projects

Environment

- Environmental management in navigation
- Climate change and emissions, energy efficiency, international regulations, carbon markets
- Fresh water availability for operations
- New technologies on infrastructure, pollution prevention, port reception facilities and ballast water
- Multiple purpose water resource systems (transport, energy, recreation, ecosystems, watersheds, potable water, ...)
- Navigation as a green transportation mode
- Societal awareness and responsibility, combining economic growth, environment / sustainability and welfare
- Climate Change Mitigation in the Maritime Industry

All these abstracts represent an invaluable source of information, knowledge, experience, and ideas useful for students, engineers, researchers, scientists, authorities and for everyone interested in the fields covered in this World Congress.

Name________________________________
Title__________________________________
Position_______________________________
Affiliation_____________________________

Le Havre - Port 2000 : A new Containerport with a simultaneous move towards environmental rehabilitation of the Seine Estuary (1996–2016)

by

Pascal Galichon[1] and Paul Scherrer[2]

1. Presentation of the project

. The "Port 2000" container facility was studied at the end of the 1990's, built starting 2001, and had its first commercial operations in April 2006 (Figure 1).

It was born from the will of the French State to position Le Havre as a main gateway for the flows of containerised goods. Today, it has 3500 m of heavy duty container quays for vessels of 16+ meters of draught.

Figure 1. Port of Le Havre after Port 2000 expansion

Total costs of Port 2000 port works have been approximately 900 Million € (or MM €) public money plus 600 MM € private money for all terminal facilities

Today, the port accommodates every days of the year, in any tidal condition, ships of more than 10 000TEU's and often up to 20 000 TEU's

[1] Director in charge of Sustainable development at Port of Le Havre Authority. pascal.galichon@havre-port.fr

[2] Former Deputy General Manager and P2000 Project Director in Port of Le Havre Authority, PIANC First Delegate for France and French representative to ENVICOM. pscherrer76@gmail.com

The expansion also provided the opportunity to contribute to the environmental restoration of the Seine Estuary (Figure 2) with a budget of approximately 50 MM €.

To develop new intertidal wetland (mudflat) the following elements were constructed: a groin plus modifications of low-crested breakwaters plus the dredging of a purely environmental channel (> 1,5 MM m^3) (21 MM €).

Two bird resting areas also were constructed, one on land using existing hunting ponds (3 MM €) and one in the south of Seine Estuary with the creation of an artificial island made of sand and gravel from the port works (9 MM €).

Figure 2. Habitat improvements in the Seine Estuary showing the dredged environmental channel .

Beyond these efforts, an important survey program was performed to increase the existing knowledge of the environment of the Estuary and the Port Authority contribution to the management of "Nature Reserve of the Seine Estuary" increased.

2. Port 2000: Applying the Working with Nature Philosophy

Step 1: Establish project needs and objectives

Port of Le Havre is at the mouth of the Seine River, at the entrance of English Channel and is a natural gateway to Normandy, Paris and North-West Europe. The Port 2000 vision was to leverage this nautical position to reinforce Le Havre as one of the major entrances to Europe for all sizes of containerships arriving from Asia or the Americas. The objectives of the port project included guaranty of quality and safety for the ships and goods entering Europe, competitiveness of port operations, increased employment for the region and increased capacity to accommodate long term growth of container traffic and size of ships.

The discussions with stakeholders pointed out that the weakest part of river Seine outer estuary were wetlands. Therefore the environmental objectives of the project focused on wetland restoration and creation. Wetlands (including mudflats) are much valued in the Seine Estuary due to historical losses, making the wetland component of the project integral to its success.

Step 2: Understand the environment

In the 1990's, during the project's initial development, global environmental studies of the Seine Estuary, were performed to include the measurement of currents wave patterns and sedimentological processes. These studies built on previous studies performed in preparation of the Bridge of Normandy construction. Studies included measurements in the field, and physical and mathematical modeling to support an understanding of hydrodynamics and sediment transport in the river.

Biological studies also were performed and included studies of fishes and fish nurseries; a comprehensive survey on bird habitat; a study of the use of the estuary by birds, permanent and migratory; and a survey of amphibians and plants. The studies were performed in the port area, and to understand the larger estuary conditions outside and upstream of the port.

These studies were analyzed and summarized in a report prepared by a "Committee of Independent Experts on the river Seine Estuary" (1999). This Committee also presented recommendations for the development of the environmental and habitat components of Port 2000 project.

Step 3: Make meaningful use of stakeholder engagement to identify possible win-win opportunities

The Port Authority decided to engage stakeholders and their consultations at an early stage to facilitate dialogue, mutual understanding and public acceptation of the project. This approach helped avoid unnecessary "stops and gos" in the project associated with public concerns.

There were many informal discussions with multiple stakeholders as early as 1996, well before the four months of official Public Hearings (Débat Public) were held in 1997 and 1998. This consultation of stakeholders continued through all phases of work, through construction, and up to the start of operation in 2006.

Thanks to all those discussions with stakeholders, the importance of wetlands and mudflats in the Seine outer estuary was put forward and it was decided to concentrate many of the environmental measures on these wetlands. It is also after dialogue with stakeholders that the building and also the design of a bird island in the south of the estuary was decided.

Specific attention was given to fishermen interests as in early years, fishermen associations opposed the project due to concerns of impact to fish nurseries in the Seine River and Estuary, and also due to concerns regarding reduced fishing opportunities due to port operations and the increased turbidity.

Results of all studies were shared, thus building relationships among stakeholders year after year. Port engineers worked with fishermen associations to address their concerns, build trust, and protect the environment. During construction, before any new phase of construction, in particular before dredging, meetings were held with fishermen and contractors to help protect fisheries and the environment to the maximum extent practicable.

Step 4: Prepare initial project proposal/design to benefit navigation and nature

The design of the groin and environmental channel for mudflat creation used the physical and mathematical models developed for the port design, to identify and realize opportunities to minimize sediment transport, optimize habitat creation, and maximize habitat stability. These measured led to win-win-win approaches that benefited Port 2000 economics, public perception and the environment.

The extensive hydrologic and sediment transport studies conducted at the beginning of the project showed that one of the risks associated with the Port 2000 expansion was the threat of upstream (eastward) sediment transport due to the acceleration of currents from Seine Bay during flood tides. These sediments could have moved towards existing mudflats with adverse effects including burial and smothering of existing habitat. To address this risk, the project included morphological dredging of approximately 3.5 MM m^3, south of the breakwater, to remove sediment with the greatest potential for upstream transport from the estuarine system.

Regarding the main breakwater of 5 kilometres an innovative design was decided with first building of a +3m (marine-level) sub-base made with dredged gravel (Figure 3). This substantial re-use of dredged material was cost effective and provided environmental benefits by minimizing dredged sediment disposal.

Figure 3. Breakwater sub-base made of gravel

The necessity of a bird island in the southern part of the estuary emerged from stakeholders during the Public Hearings. This island (Figure 4) was designed with a fruitful dialogue with ornithologists and constructed south of the estuary, also using dredged material from the port construction.

Figure 4. Bird Island constructed using dredged sediment.

Globally the project reused 26 MM m^3 sediment, of the total 45 MM m^3 dredged for the new channel and basin.

Also it is good to point out the plant survey demonstrated the presence of a very rare and protected orchid named "*Liparis loeselii*". As a consequence the port modified the project to exclude the area favorable for L. loeselii, thus minimizing unintended habitat impacts.

Stage 5: Build and implement

Construction of the project also considered environmental impacts and established measures to minimize environmental harm. Contractors were required to use mathematical models (Figure 5) to phase aquatic construction (e.g., dredging, building the breakwater) to identify progressive solutions that recognized current velocities and sediment transport potential.

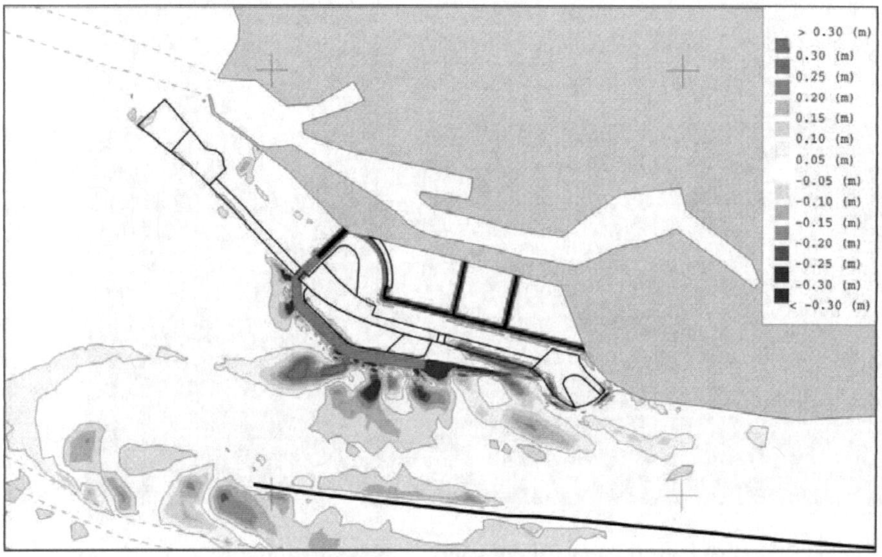

Figure 5. Mathematical modelling of Breakwater

Physical and mathematical models also were used to evaluate the stability of the gravel sub-base of breakwaters in the presence of natural currents, with the goal of working with and not against natural water flows, thus achieving substantial economic improvements.

Building of the groin and the environmental channel was performed adaptively, to allow adaptations to the design during construction, in accordance with actual sediment transport processes.

Stage 6: Monitor, evaluate and adapt

The Port Authority established a 10+ year monitoring program that extends from the Bridge of Tancarville to the sea. The monitoring program aim and objectives were defined by an independent scientific committee active in Seine estuary, before it was officially approved by local state government (Préfet de Région) Regular presentations of monitoring results have been made to this scientific committee that has been an active participant in the monitoring, sometime asking for additional surveys.

The monitoring program includes specific attention to fish, birds and amphibians. Monitoring also extends to sediments, water quality, and all type of species living in the estuary. For fishes, monitoring also includes a socio-economic evaluation of the fishing industry to give assurance to fishermen that the new port does not adversely impact fishing.

The Port Authority shares its experiences with the public and with other port authorities to transfer knowledge and expand their experiences universally. Beyond meetings with stakeholders, which are still ongoing, the Port of Le Havre organized and led an International Symposium in May 2015, focused on Port 2000 environmental measures, in combination with the "Estuarine Coastal Sciences Association."

For any additional information contact Paul Scherrer the project manager of Port 2000 from 1994 to 2014 (pscherrer76@gmail.com) or Pascal Galichon, the deputy project manager of Port 2000 and current Manager of Sustainable Development at Port of Le Havre Authority (pascal.galichon@havre-port.fr)

REFERENCES

CNDP (Commission Nationale du Débat Public) 2002 – 2012 LE CAHIER DE MÉTHODOLOGIE https://www.debatpublic.fr/file/350/download?token=eJgaTugW

GREENPORT (2015) France's HAROPA Ports has won a best green seaport award voted for by 18,000 Asian shipping and logistics professionals http://www.greenport.com/news101/europe/best-green-seaport-accolade

PIANC Working Group 176 report (2018) : Guide for Applying Working with Nature to Navigation Infrastructure Projects

Developing a method to grasp coral reefs through remote sensing technology

by

Etsujiro KATAYAMA[1], Toshiaki KOMATSU[1], Yasuo YONEZAWA[1], Mika KATAYAMA[1], Toru KANYAMA[2], Masayuki FUDO[3]

Abstract

Due to the effects of land reclamation for harbors, etc. and development activities such as sea route dredging, as well as the global influence of an increase in the seawater temperature, etc., coastal coral reef sea areas around the world are exposed to risks such as bleaching and coral disappearance, which may be accompanied by a damaging effect on the diverse living things inhabiting the coral reef, the fisheries industry, sightseeing and the disaster prevention functions of the coral reef. For Madreporaria alone, which is the main reef-building coral, there are 839 species on the IUCN red list. For that reason, in the case that development activities are conducted in coral reef sea areas, it is important to evaluate the impact, and to consider avoidance, minimization and corrective measures with regard to the impact. Here, using two types of remote sensing technology, a means of accurately understanding the coral distribution in coral reef sea areas will be introduced. The first is (1) a technology for understanding coral distribution using high resolution multispectral satellite imagery, and the other is (2) a technology for performing a 3D analysis of coral reef and understanding coral distribution using underwater video footage. (1) is able to accurately understand and analyze wide-area coral distribution and coverage in coral reef sea areas at a scale of tens of km^2. (2) is able to specifically understand and analyze the terrain of coral reef in an area extending approximately ten meters in every direction, and the distribution of individual corals and seaweed. As an example, we will report the results taken from Okinotorishima in Japan. Okinotorishima is a shallow island measuring approximately 4.5km east-to-west, 1.7km north-to-south and with a depth of less than 10m. For the understanding of coral distribution from the satellite imagery in (1), it was possible to analyze the coral distribution and coverage with 80% classification accuracy as a result of obtaining satellite imagery of the entire island and applying coral coverage guidance data based on image clustering and field surveys. By obtaining satellite imagery for several years and conducting the same analysis, we were able to understand the change in the coverage area over the years. For the underwater video footage analysis in (2), video footage was taken of knolls in the coral reef in an area extending ten or more meters in every direction, and we were able to make a three-dimensional understanding of the coral distribution using a 3D analysis. Also, as a result of classifying the coral in a part of this area, classification accuracy of 80% or more was obtained. These technologies can be considered as useful tools to quantitatively understand the influence of development activities such as sea route dredging and the effects of prevention measures such as coral transplanting.

Keyword: multispectral high resolution satellite imagery, underwater photography, coral reef, coral classification

1. Method

1.1 Satellite imagery coral classification process

In this study, high resolution satellite imagery IKONOS GeoEye-1 (resolution of 0.5m) was used in the multispectral satellite imagery. The years of photography were 2006, 2011, 2012 and 2017, which are recent years in which there is little noise due to clouds and high waves, etc.

The flow of coral coverage analysis using satellite imagery is shown in Figure 1.

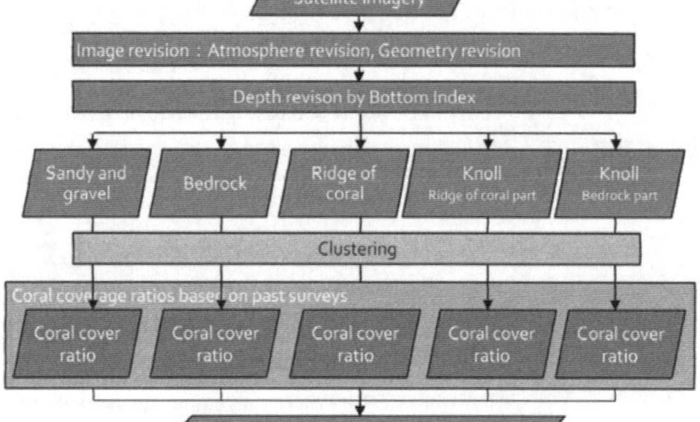

Figure1 Satellite imagery coral classification flow

[1] Environmental Engineering section, KOKUSAI KOGYO Co, Ltd., Japan, etsujiro_katayama@kk-grp.jp
[2] Fisheries Infrastructure Development Center, Japan
[3] Fisheries Agency, Japan

First, for pre-processing of the satellite imagery, atmospheric correction using dark pixel subtraction and geometric correction using geographic coordinates as reference points were performed. Then, the bottom index was calculated, and the calculated bottom index image was overlapped with the satellite imagery to divide the image into individual bottom quality divisions using existing bottom quality divisions. Unsupervised classification (clustering) was performed on the image after sectioning off the bottom quality divisions. A coral coverage distribution map for each class after classification was produced by applying the coral coverage after overlapping the existing coral study results with the 2006 satellite imagery taken at the same time. Regarding the satellite imagery classification results from other years (2011, 2012 and 2018), the coverage was automatically applied using the relationship between the image data from each class according to the 2006 analysis results and the coral coverage.

1.2 3D analysis and coral distribution analysis process using underwater video

The flow of the investigation and analysis is shown in Figure 2. In the field survey, divers took underwater video footage in the survey target area. In this study, the photographs were overlapped in order to prevent any gaps in the images of the knolls in Okinotorishima in an area extending about ten meters in every direction. Two types of photography processing were performed, namely, mosaic-type and orthographic-type using 3D imaging. The mosaic type forms a plane figure for the entire survey target area by pasting together the characteristic points of the overlapping images in the divided background images. 3D image processing forms a 3D shape using SfM (structure from motion). Then, from the produced 3D model, an orthographic projection is made on any surface to produce an orthographic image. Finally, for the automatic classification of coral, coral distribution is confirmed using object base classification. In this study, there were 2 classifications (coral and non-coral) and 7 classifications (branch-like coral, other coral, dead coral, seaweed, bedrock, giant clam and other).

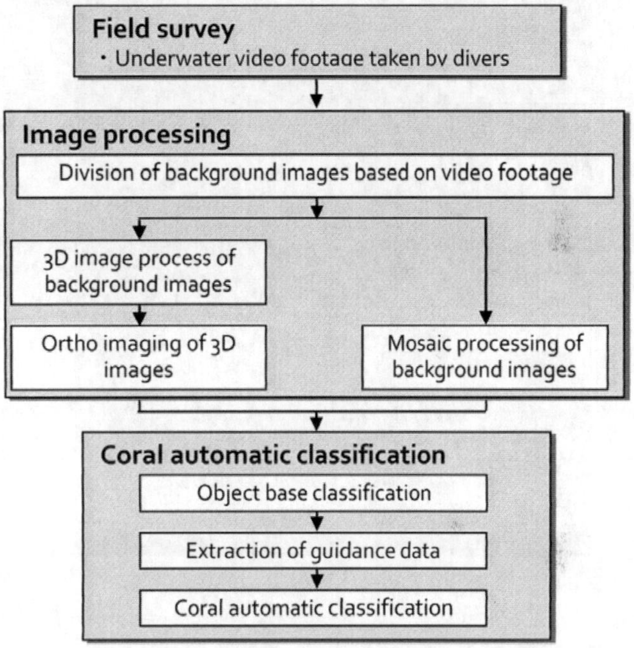

Figure 2 Underwater video footage analysis

2. Results and Discussion

2.1 Satellite imagery coral classification results

The coral coverage distribution map and the transition in coral area calculated therefrom are shown in Figure 3. Based on this analysis, in the approximately 11 years between 2006 and 2017, it was confirmed that the coral area reduced by roughly half.

A verification of the accuracy of the multispectral satellite imagery analysis was performed using the results of the analysis of 2012 satellite imagery and the 2013 to 2014 field survey results. The image classification accuracy verification results are shown in Figure 4, Figure 5, Figure 6 and Table 1.

As for the accuracy of the coral coverage analysis results based on satellite imagery in contrast with field verification data, the accuracy in an RMSE (root mean square error) evaluation was approximately ±3%, and the overall accuracy was 80% with a Kappa coefficient of 0.7 (substantial).

The cost of performing the coral distribution analysis was, in the case of an analysis area of 10km^2, approximately 20,000 dollars for coral classification by satellite imagery, whereas the cost in order to obtain the same accuracy from a visual inspection performed by divers is calculated to be 200,000 dollars, so an understanding can be gained with about 1/10th of the cost.

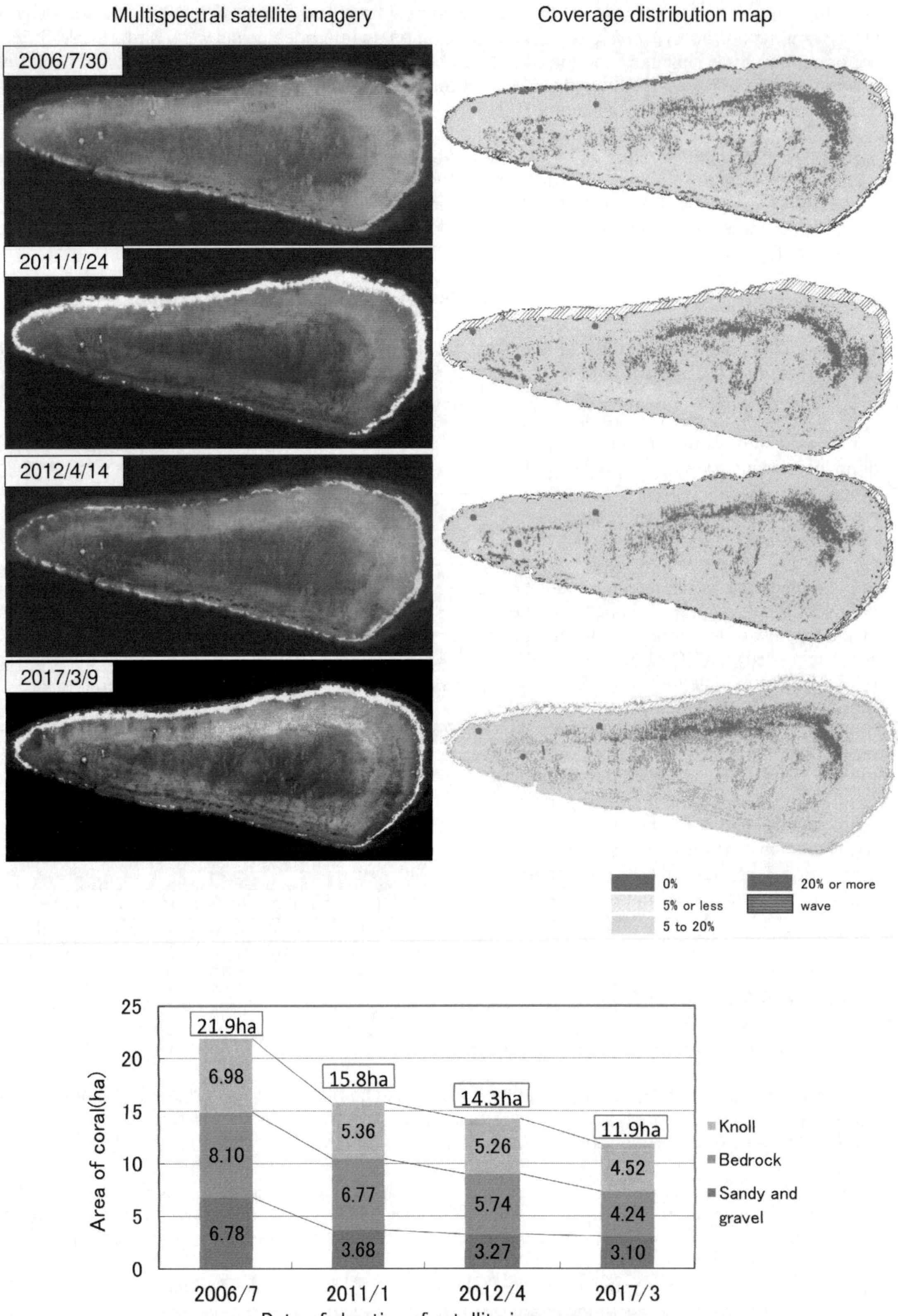

Figure 3 Coral coverage analysis results (example of Okinotorishima)

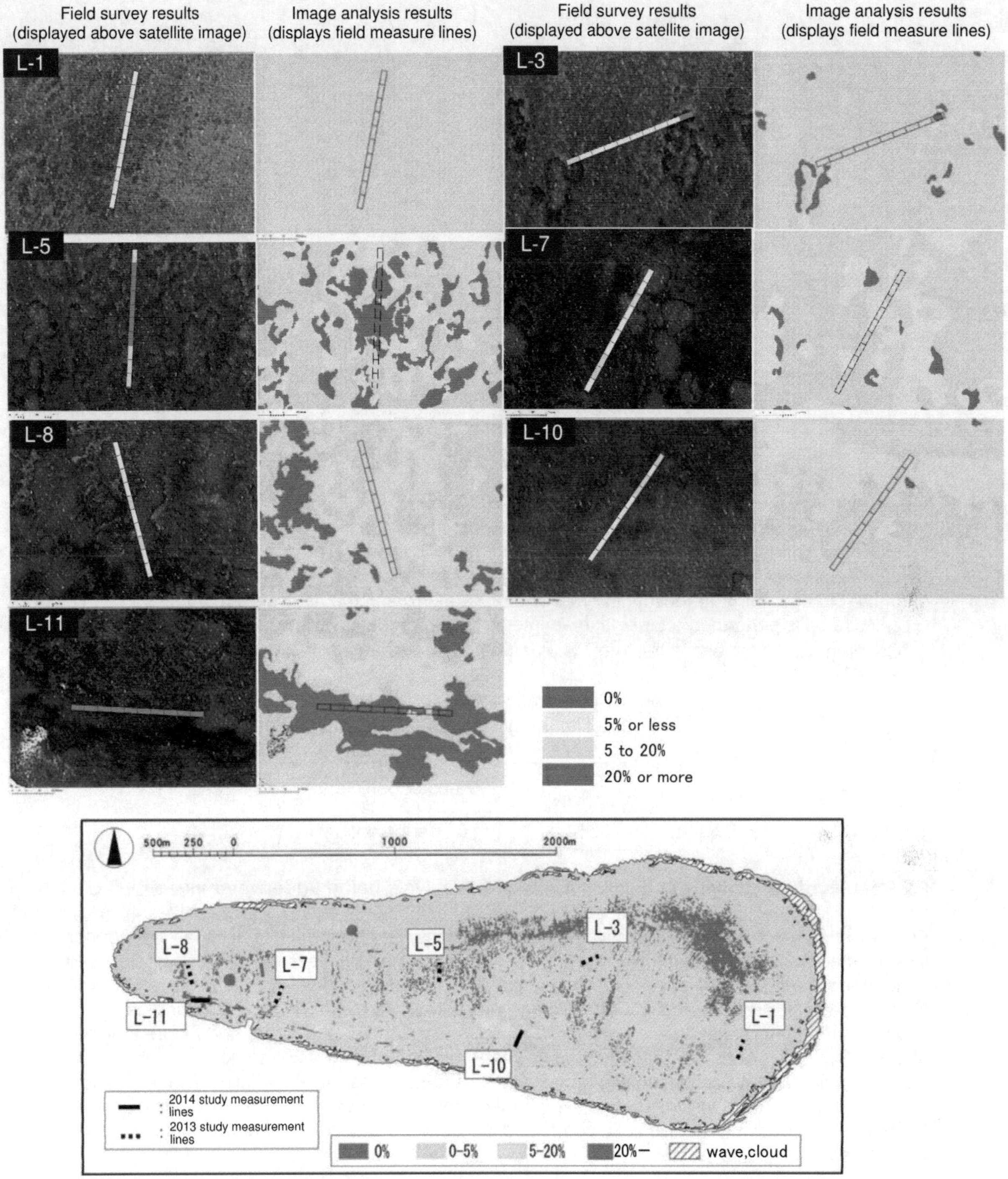

Figure 4 Coral coverage field survey results and comparison with satellite imagery analysis results (Field survey: 2013 and 2014; Satellite image: 2012)

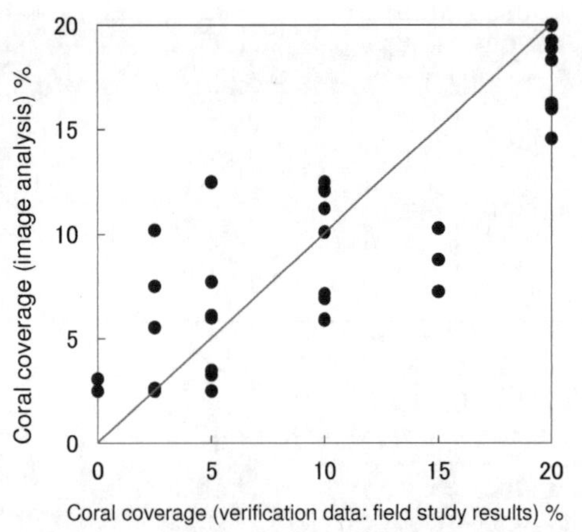

Figure 5 Comparison between coral coverage verification data and image analysis results (2012 satellite imagery, 2013/2014 field surveys)

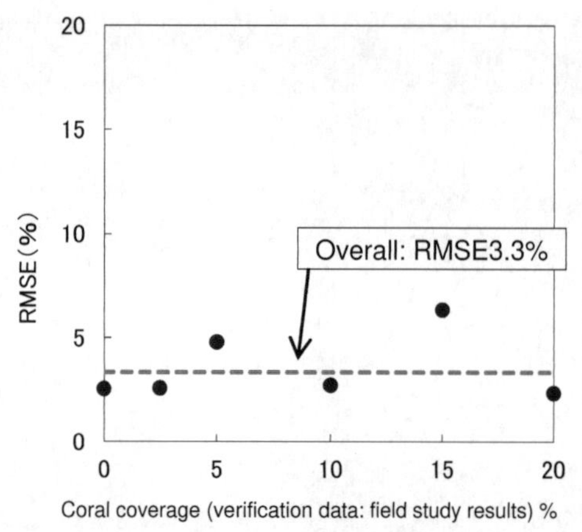

Figure 6 Calculation errors in image analysis results (RMSE) (2012 satellite imagery, 2013/2014 field surveys)

*1. Coral coverage based on image analysis shows the total value (average) of the coverage information in all images (40 images) in a 4m x 10m observation area.
*2. Coral coverage based on a field survey and image analysis is the data needed to show the scope of the change. For that reason, in order to calculate errors in the analysis, a comparison and error calculation is performed after setting an integer value for the coverage, as shown below.
Coverage of 1% or less⇒0%, 1 to 5% or less⇒2.5%, 5 to 20%⇒12.5%, 20% or more⇒20%

Table 1 Classification accuracy verification results (2012 satellite imagery, 2013/2014 field surveys)

Overall accuracy	Kappa coefficient
80%	0.7 (substantial)

2.2 Results of 3D analysis and coral distribution analysis using underwater video

Figure 7 shows examples of coral coverage classifications after performing a 3D analysis of underwater video filmed in Okinotorishima. As the Figure shows, by performing a 3D analysis, it is possible to understand the classification of coral in specific parts of knolls, as well as their size and distribution. Next, a coral classification analysis of the image was performed by making it an ortho image. The results of verifying the interpretation accuracy of these images are shown in Table 2 and Figure 8. In the case of using 2 classifications, for both the mosaic image and the ortho image, an accuracy of 80% or more for automatic coral classification was obtained. On the other hand, when using 7 classifications, the accuracy of automatic coral classification with the ortho image was 40-60%, so issues were seen in terms of practicality. The causes of incorrect classifications are unclear images, and misunderstanding dead corals and scarce corals, etc.

Therefore, from now on, it is considered to be important to give attention to acquiring images that facilitate classification by means of closeup photography and the use of lighting when photographing shaded areas, for example. Furthermore, it is considered to be important that the cost of analysis is reduced by creating a model to obtain the same analysis accuracy from multiple images rather than analyzing each image, as we did in this study.

The cost of coral distribution analysis, in that case that the scope of analysis covers 1ha, the cost of coral classification analysis using underwater video is approximately 11,000 dollars, whereas the cost in order to obtain the same accuracy from a visual inspection performed by divers is calculated to be 110,000 dollars, so an understanding can be gained with about 1/10th of the cost.

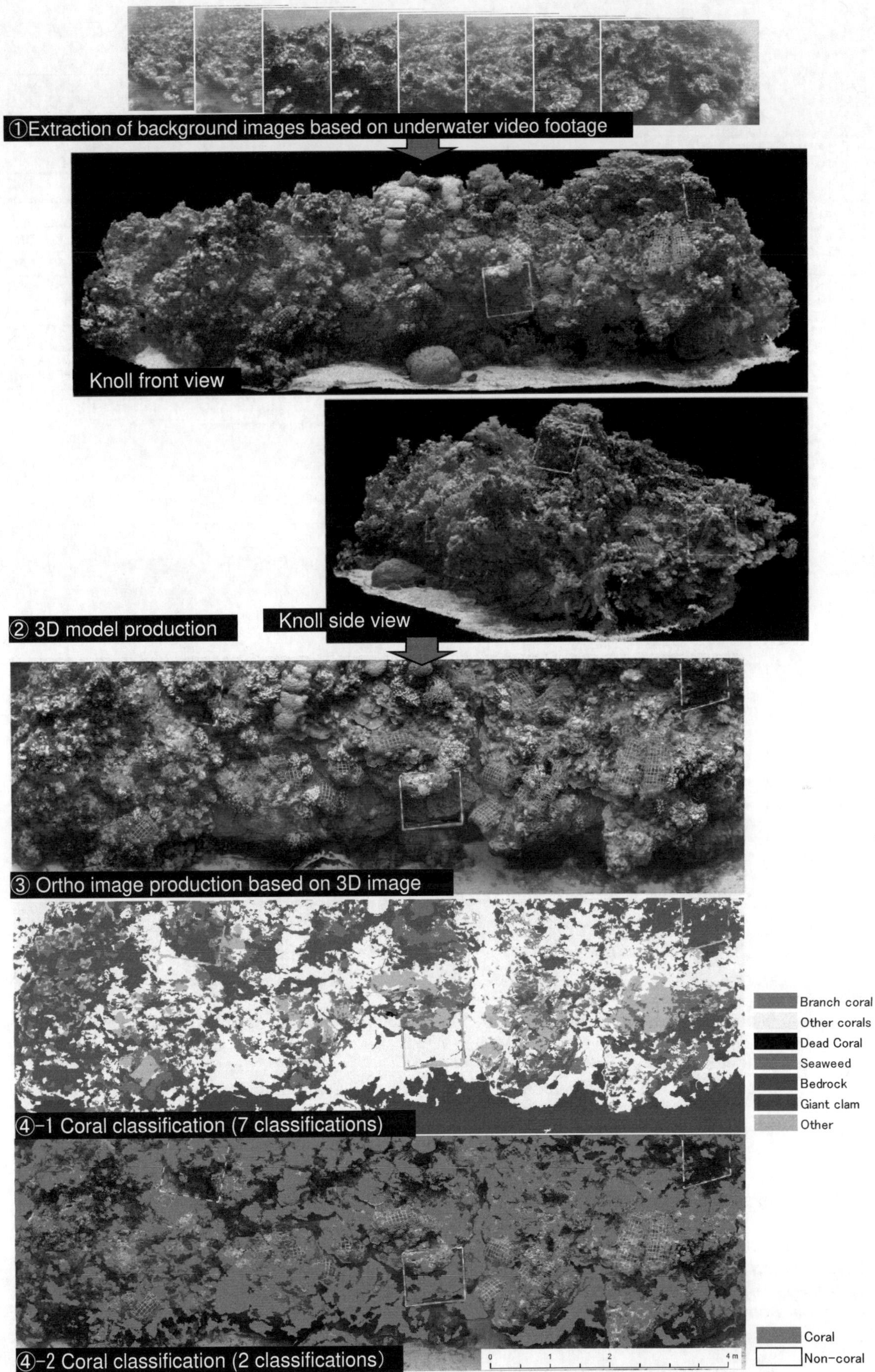

Figure 7 3D model and ortho image production, and coral classification examples

Table 2 Accuracy verification results

No.	Image	Image scale	Guidance data	Guidance classification item	Year and location	Automatic classification accuracy	
						2 classifications	7 classifications
1	Mosaic imaging	Approx. 30m^2	Applied to each image	Color, pattern, scale, etc. (20 types)	Measurement line L-5 FY2015 Seabed Complete image	80%	-
2-1	Orthographic imaging	Approx. 50m^2	Applied to each image	Addition of depth conditions gained from orthography in color, pattern, scale, etc. (23 types)	A3_1 FY2016 Knoll vertical plane Complete image	85%	68%
2-2		Approx. 30m^2			A3_2 FY2016 Knoll vertical plane Complete image	87%	68%
2-3		Approx. 50m^2			A10_1 FY2016 Knoll vertical plane Complete image	72%	60%
2-4		Approx. 20m^2			A10_2 FY2016 Knoll vertical plane Complete image	76%	57%
3-1	Orthographic imaging	Approx. 50m^2	Applied to each image	Addition of color and pattern conditions (47 types)	A10 FY2017 Knoll vertical plane Complete image	87%	55%
3-2					A10 FY2017 Knoll vertical plane Sections with large amounts of seaweed	91%	63%
3-3					A10 FY2017 Knoll vertical plane Sections with large amounts of dead coral	84%	44%
3-4					A10 FY2017 Knoll vertical plane Sections with large amounts of bedrock	85%	63%
3-5				Same as No. 2 (23 types)	A10 FY2017 Knoll vertical plane Complete image	-	51%

*Accuracy colors: Stronger blue color denotes higher accuracy, Stronger red color denotes lower accuracy

*2 Classifications: Coral, non-coral; 7 Classifications: Branch-like coral, other coral, dead coral, seaweed, bedrock, giant clam, other

*Automatic classification accuracy = Area correctly classified by means of automatic classification/total area

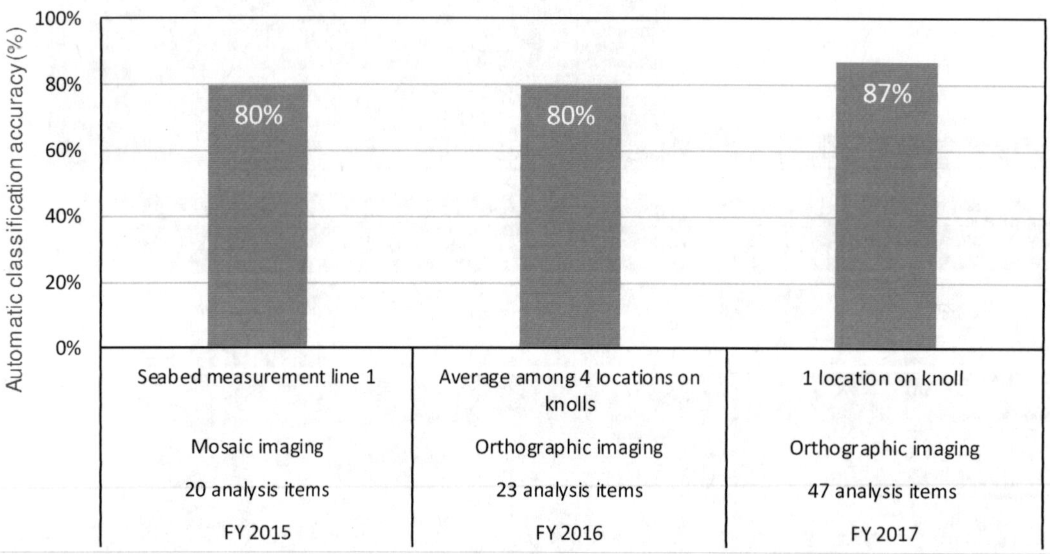

Figure 8 Comparison between coverage survey results and satellite imagery analysis results

References

Mika Katayama, Taichi Morita, Kumiko Suzuki, Yasuo Yonezawa, Etsujiro Katayama, Hiroya Yamano, Wataru Ando, Takayuki Nishizaki, Norihito Watanabe (2014). Coral distribution mapping of Okinotorishima using time-series high resolution satellite imagery for expanding the distribution of coral. Proceedings of the 56th Conference of the Remote Sensing Society of Japan, 93-94

Mika Katayama, Taichi Morita, Kumiko Suzuki, Yasuo Yonezawa, Etsujiro Katayama, Hiroya Yamano, Wataru Ando, Takayuki Nishizaki, Norihito Watanabe (2014). Characteristics of coral distribution change of Okinortorishima using time-series high resolution satellite imagery. Proceedings of the 57th Conference of the Remote Sensing Society of Japan, 107-108

Yasuo Yonezawa, Etsujiro Katayama, Mika Katayama, Toshiaki Komatsu, Wataru Ando, Satoshi Uchida, Kousuke Nakamura: Coral distribution mapping in okinotorishima by time-series high resolution satellite imagery. 13th International coral reef symposium abstracts, Hawaii, U.S.A., 2016

Etsujiro Katayama, Kenichi Honda, Yasuo Yonezawa, Wataru Ando, Kunitomo Asakura, Masayuki Fudo (2016): Coral coverage analysis using underwater video footage. Proceedings of the 19th Annual Conference of the Japanese Coral Reef Society, 20

DEVELOPMENT OF CORAL REEF PROPAGATION TECHNOLOGY BY SEXUAL REPRODUCTION, IN JAPAN

by

Masayuki FUDO[1], *Keiichi TAMURA*[2], *Takuto TSUKAMOTO*[2], *Wataru OKADA*[2]
and Wataru ANDO[3]

ABSTRACT

The coral reef is an area of primary production that cultivate ecosystems with high biodiversity, and it is also an important area as fishery resources. However, many of the coral reefs around the world tend to decline because of rising seawater temperature caused by global warming (Carpenter et al. (2008)). Meanwhile, the upsizing of ships has become a global trend along with the increase in demand for logistics, so dredging of routes and construction of harbors and fish ports are increasingly important. As a result, coral reef area came to be directly changed and coral growth came to be indirectly influenced by the coastal area development. For these reasons, in Japan, as a measure to avoid, reduce or compensate for the influence on ecosystems in the development of coral reef, we have been developing coral reef propagation technology by sexual reproduction. In this technology, it is possible to produce genetically diverse corals which can propagate themselves and supply coral larvae. So, it is considered to be effective for restoration of a wide range of coral reefs. It is composed of three technologies, which are coral production, coral cultivation on nurseries, and coral transplantation. In addition, coral production by sexual reproduction is divided into two methods; one is producing corals in water tanks on land, and the other is producing corals by collecting, rearing, and seeding natural coral larvae in situ. Here we introduce the efforts of coral propagation technology and report on the result.

Keyword: coral reef propagation technology by sexual reproduction, coral production on land, collecting coral larvae in situ, coral cultivation on nurseries, coral transplantation

1. THE CORAL PROPAGATION TECHNOLOGY BY SEXUAL REPRODUCTION

The coral propagation technology is composed of three steps, which are coral production, coral cultivation on nurseries and coral transplantation. In addition, coral production by sexual reproduction is divided into two methods; one is producing juvenile corals in water tanks on land, and the other is producing juvenile corals in situ by collecting, rearing, and seeding natural coral larvae. It is important to use an appropriate method according to the conditions of the sea.

CORAL PRODUCTION	CORAL CULTIVATION ON NURSERIES	CORAL TRANSPLANTATION
Collecting adult corals in the sea and making them spawning in water tanks on land	Outplanting juvenile corals and growing them on nurseries (until they become about 5cm)	Transplanting grown corals on nurseries to places where is suitable for coral growth
Collecting, rearing, and seeding natural coral larvae in situ (using "Larval Cradle")		

Figure 1: The flowchart of the coral propagation technology by sexual reproduction

[1] Fisheries Agency, Japan
[2] ECOH Corporation, Japan
[3] Fisheries Infrastructure Development Center, Japan

Figure 2: The process of the coral propagation technology by sexual reproduction

2. CORAL PRODUCTION BY SEXUAL REPRODUCTION

2.1 Mass Production of Corals by Sexual Reproduction on Land

Our development of coral propagation technology began with producing a large number of juvenile corals. by sexual reproduction. The target species was *Acropora tenuis*. We collected adult corals in the sea, which were used for the spawning in large water tanks on land located at Okinawa. In the period from 2006 to 2008, we succeeded in producing about 30,000 juvenile corals (Nakamura et al. (2011)).

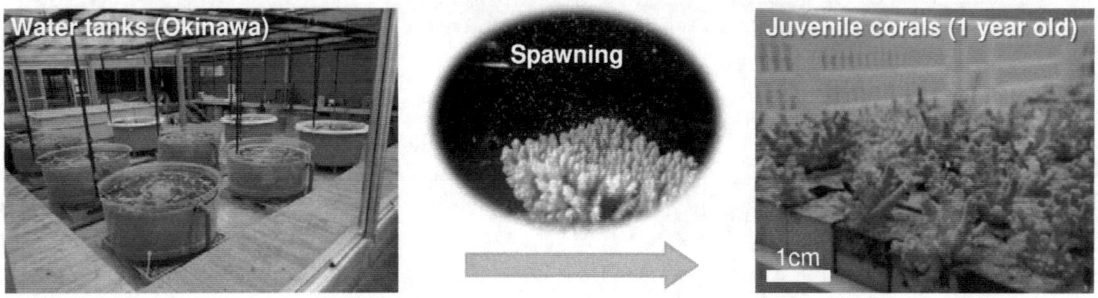

Figure 3: Coral production in water tanks

2.2 Collecting, Rearing, and Seeding Natural Coral Larvae in Situ

While producing a large number of juvenile corals in water tanks on land, we are currently undertaking the development of technology that can produce them in the sea. For that purpose, we developed an equipment that is capable of collecting, rearing, and seeding natural coral larvae in situ.

We call the equipment "Larval Cradle". The shape of it is cylindrical to reduce resistance to water flow. It is made of the nylon net, and appropriate mesh size of it, an important factor for enhancing survival rate of the enclosed larvae, is 30 μm. The maximum number of coral larvae which can be collected by larval cradle was 3 million at 4 days-old, which they can settle.

With "Larval Cradle", it is possible to produce corals in the area where is not able to make water tanks system, and to produce mass corals at a considerably lower cost compared with the method of producing corals in water tanks on land. We confirmed that the coral larvae settled on about 6,000 settlement devices.

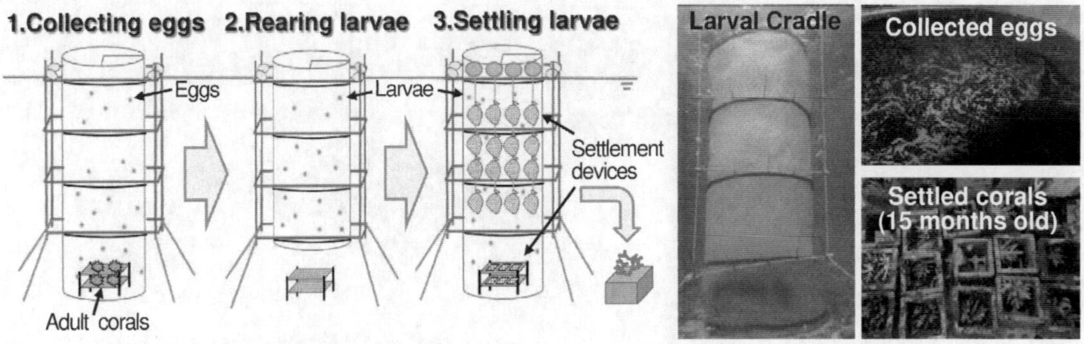

Figure 4: The image of producing corals by using Larval Cradle in the sea

3. OUTPLANTING AND GROWING MASS JUVENILE CORALS ON NURSERIES

We developed artificial structures (such as artificial concrete blocks) designed as nurseries for juvenile corals, and installed them to the unsuitable area for coral growing such as sandy bottom.

And we also developed lattice shaped foundation for efficient outplant of a large number of juvenile corals on the structure, and made it possible to efficiently outplant about 200 juvenile corals per person-day. After the outplant, we monitored and maintained them on a regular basis to keep them in good condition. Now, since the survival rate of coral one year after the outplant is as high as 80 %, we believe that our technology has been mostly established. The juvenile corals outplanted with a diameter of about 1 cm grow up to about 5 cm in one year.

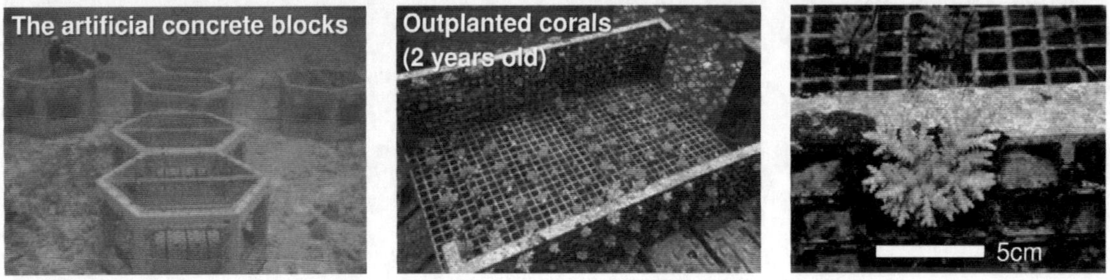

Figure 5: Corals outplanted on the artificial concrete block

4. TRANSPLANTING TO EXPAND CORAL HABITAT AREA

After choosing the suitable area for coral transplantation, we transplanted grown corals on nurseries to the area to expand coral habitat area. It is required that the transplantation area is suitable for coral growth. In addition, if transplanted corals spawn, it is preferable that coral larvae don't flow out to the open sea. Therefore, we selected suitable area for coral transplantation by utilizing field survey and numerical simulation technology.

We are trying to develop of "the Larval Supply Base". If transplanted corals which has genetically diversity have grown and matured, they supply larvae to surrounding areas autonomously. We expect that such "larval supply base" will recover extensive coral reefs in near future.

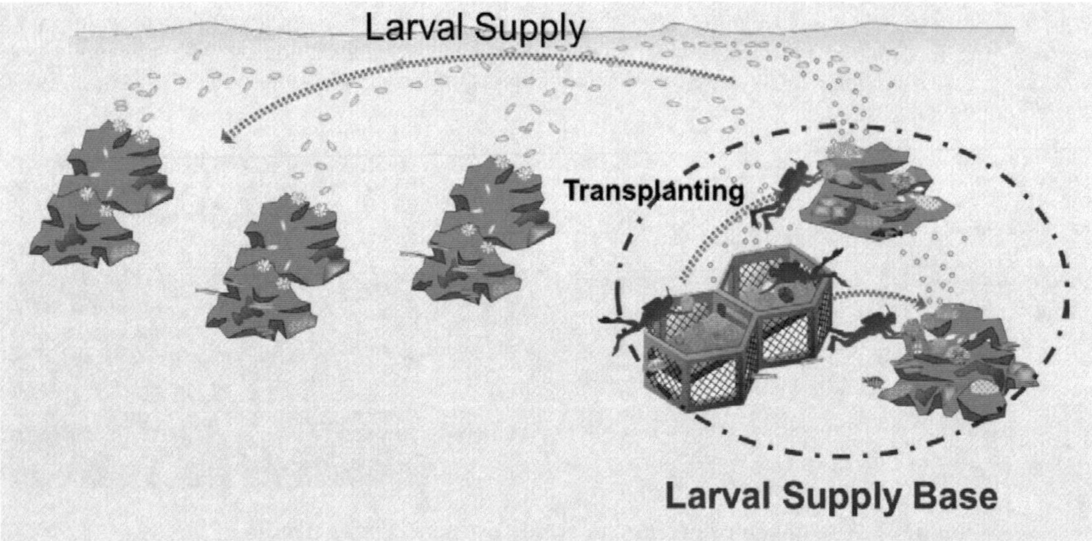

Figure 6: The schematic image of intensive coral reef restoration by larval supply base

References

Carpenter KE et al. (2008). One-third of reef-building corals face elevated extinction risk from climate change and local impacts. Science 321: 560-563.

Nakamura R, Ando W, Yamamoto H, Kitano M, Sato A, Nakamura M, Kayanne H, Omori M (2011). Corals mass-cultured from eggs and transplanted as juveniles to their native, remote coral reef. Mar Ecol Prog Ser 436. 161-168.

THE CHALLENGES OF LIMITING THE ENVIRONMENTAL IMPACT OF FAIRWAY PROJECTS

by

Camilla Anita Spansvoll[1,2]

INTRODUCTION

The Norwegian Coastal Administration (NCA) is an agency of the Norwegian Ministry of Transport and Communications responsible for services related to maritime safety, maritime infrastructure, transport planning and efficiency, and emergency response to acute pollution. NCA activities encompass the maritime sector of the National Transport Plan (NTP). The NTP outlines how the Norwegian government intends to prioritize resources within the transport sector. Further, the NCA executes authority and administrative tasks related to the laws and regulations for ports, fairways, and compulsory pilot services.

By the example of a recent projects the NCA aims to share experiences and solutions on how to avoid negative environmental impacts in the execution phase of projects for new or improved fairways. The focus in this article is underwater noise pollution and fairway pollution. The NCA present both what went well through pre-project planning and execution, what could have been performed better as well as ongoing internal projects on mitigating measures. Through this, we hope that others who faces similar challenges will be encouraged to contact the NCA for future collaboration and international experience exchange.

THE NCA PORTFOLIO – THE COAST

The coast has always been a lifeline for the Norwegian society. Today, 90 % of the Norwegian export revenue origins from activity and resources connected to the sea (1). Increasing the quality of fairways is a well-known risk reducing measure for vessel traffic. The Norwegian mainland's coastline is almost 29 000 km long, only exceeded by the Canadian coastline. By including islands, the total coastline is just over 100 000 km. The coastline has its southernmost point at 57° south and the northernmost at 71° north. This presents an environment where one must take into account a high diversity, both of biological- and geological factors, when construction projects are planned and conducted. The climate also represents seasonal challenges such as ice, cold and severe storms.

When a new fairway is developed, existing fairways are improved, or a harbor is dredged, pollution is often an issue. Industry, including i.e. fisheries, oil and gas support services, and shipyards has been traditionally, and still are, located along the coast and other large water bodies connected to the sea. For decades, it was common practice to use the sea as an easy way for waste dumping. For the Norwegian ports and fairways, we like to think of this as something that is a closed chapter when it comes to "waste management". Nevertheless, our past environmental sins become present, also for NCA projects. Deep layers of various types and degrees of pollution create challenges and severely increase the cost and complexity of remediation projects. The removing of toxins and other type of pollution through a fairway or harbor project is in most cases positive for the environment – if conducted right. Considerations must be made to prevent the spreading of both clean and polluted sediments. This also includes e.g. runoff containing toxins from a disposal site or damages on the environment by increased turbidity. Dredging of polluted harbors and the safe storage of polluted sediments, reduction in greenhouse gas emission through shortened fairways, reduction of risk of accidental spills by improving fairways are important parts of NCAs work.

---

[1] Norwegian Coastal Administration (NCA), Center for construction, camilla.spansvoll@kystverket.no

[2] Environmental advisor, NCA Nordland

THE NEW FOCUS - UNDERWATER NOISE FROM CONSTRUCTION AND EXPLOSIVES

In total 16 out of Norway's 20 counties (not including Svalbard) has a coastline. When preparing and executing projects, the NCA applies for an environmental permit at the County Governors environmental department which has the governing authority. The NCA as the states naval construction developer has the burden of proof for whether or not the construction plan, mitigating measures and considerations in regard of environmental issues are satisfactory.

Due to a lack of factual knowledge in regards to some of the challenges NCA projects faces, permits and terms are at times characterized by precautionary measures. In recent years, this has in particular been the case when explosives are used to remove bedrock. How do shock waves from seabed-blasts affect fish, marine mammals and other wildlife? There is no doubt that physical damage is the result when in close proximity, but what can be considered to be a safe distance?

Will blasting of bedrock disturb spawning areas two kilometers away from the site? If yes, what mitigating measures can be used to prevent this? These questions have resulted in a NCA-initiated project measuring underwater noise and pressure from both ramming of piles and the use of explosives.

The restrictions often connected to the noise from construction and blasts comes "on top" of other restrictions. Construction work is often not permitted to be carried out in the period 15th of May - 20th of August. This corresponds to the spring – summer breeding season for birds and other wildlife. Spawning-season for various types of fish is largely stretched between February to October. Based on local knowledge this will also be set as restrictions when work is planned in the vicinity of spawning areas or seasonal migration ways.

As the governing authorities have had an increased focus on underwater noise, the demands set for NCA projects have been colored by this. The NCA experienced that the safety zone set between a blast-site and e.g. a spawning area is very varying from permit to permit. In some projects, this will to a large degree afflict the time-window given for the construction phase.

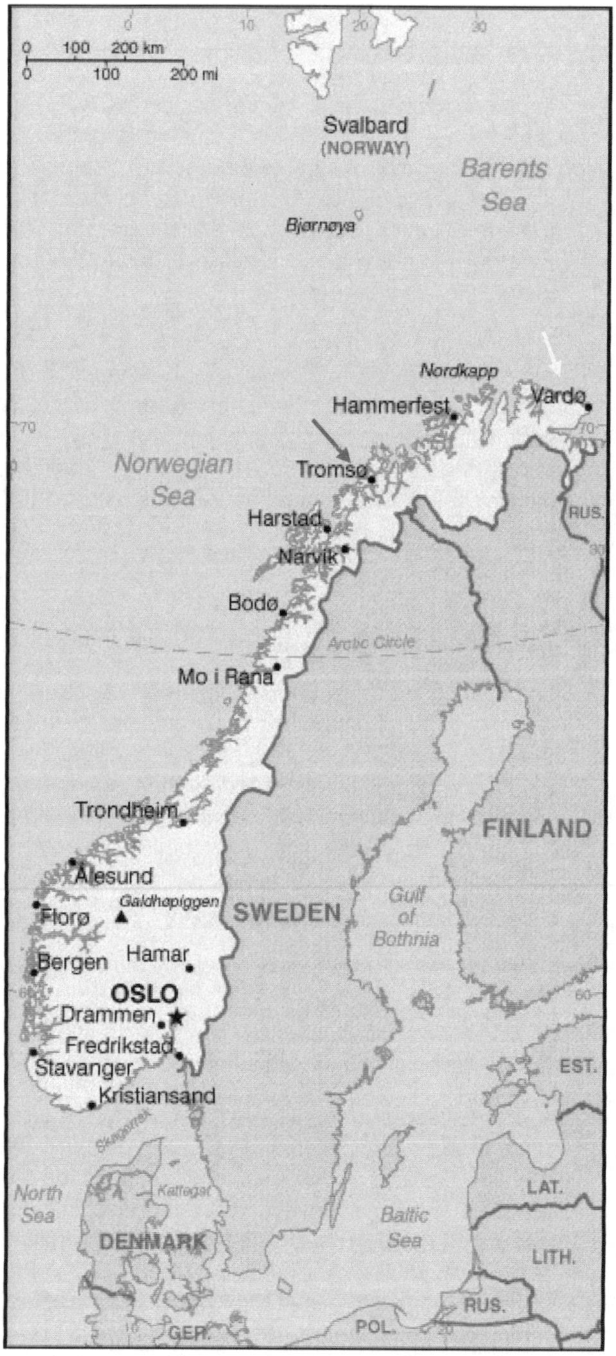

Figure 1. The Norwegian coastline. Red arrow indicates Kvalsundet, yellow arrow Baatsfjord (map: wikipedia.org).

Case-study - Kvalsundet

The fairway project though Kvalsundet (**Figure 1**) is a case that got its fair share of media attention at the end of 2016, with the majority of this being negative and based on the public not understanding the project and the way the construction was conducted. The Kvalsundet project was triggered by the need of a safer and shorter fairway into the city of Tromsoe. The existing fairway was limited for large ships by the height of bridges connecting the island, where Tromsoe is located, to the mainland and neighboring islands. By removing shallows and broadening the fairway through Kvalsundet, the detour many large ships had to make to avoid bridges was reduced with 80 km when northbound. If Tromsoe was the final destination before returning south, the ships will now be spared a sailing distance of 160 km. The new fairway was also upgraded with new and better navigational marks, significantly improving safety and reducing the risk of collisions or foundering.

The fairway project was originally planned to be finished by the 1st of October 2016, but due to delays the construction work was by this date expected to pass the original date by several months. The main concern for the NCA was that by the end of November, the waters in the surrounding area are packed with winter migrating herring and a great number of whales feeding on the herring (**Picture *1***).

The NCA and the County Governors environmental department discussed the case when it became apparent what was happening. The migrating herring's had visited this area at that time of the year for a couple of years, and it was decided by the NCA that the construction work had to be paused. The ongoing work at that time was the removing of shallows to broaden the fairway. Explosives were used for this purpose, and there was no data to be found on the potential impact of noise and shockwaves inflicted on herring and whales. Both fisheries and tourism followed in the wake of the herring and the whales. Fishermen were worried that their livelihood could be damaged, tourist providers that the whales would leave. The risk was too great, both with uncertainty of the potential damage the blast could bring and the potential negative attention the continuation of the project would bring through media and social media. Therefore, the work was completed during early spring 2017. In hindsight, the NCA could have foreseen the possible complication the contractors delay would cause, and this will be taken into consideration in future projects.

Picture 1. Pod of orcas in Kvalsundet (photo Lene Gjelsvik, NCA).

By the end of September 2017 concern was raised again, this time regarding the delay in constructing navigation installations. The project was not completed, and it was a possibility that the herring and whales could migrate into the same area again. The remaining work by this time on the navigational marks (tripods) was the ramming of piles into the bedrock (**Picture *2***). The governing authorities were not concerned that this work could impact fish and wildlife in a negative way, but the NCA wanted to be prepared – just in case. It was decided that the noise from the ramming was to be monitored and measured, and a surveillance program was implemented that the contractor had to follow. The work needed to be stopped immediately if herring and whales were observed and predefined measures on how to proceed were given.

However, the herring winter migration for 2017 changed from the previous year. The schools turned up much further north – and stayed there for the extent of the period. This also included the great number of whales following in the wake of the herring. Nevertheless, the already planned measurements were performed in Kvalsundet. The value of this work might not be crucial at that time, but it was clear that it would be important to collect the data for future projects.

Picture 2. Positioning of pile - Kvalsundet (photo: NCA).

Methods and preliminary results

Measurements from piling

During the 21st and 22nd of November, four recordings of the ramming of two piles were conducted. This included a total of data of 580 hammer strikes (**Figure *2***). As the analysis of the measurements of the piling work is not complete at the present time, a conclusion cannot be presented in this paper. However, preliminary results show that the frequencies that were recorded were found to be much higher than expected.

New case study - measurements from blasting bedrock and test of mitigating measures

The NCA also has an ongoing fairway-project in Baatsfjord, Finnmark County (**Figure 1**, indicated by yellow arrow). Bedrock is removed to increase sailing depth in several parts of the harbor (**Figure *3***). Underwater measurements of noise and sound pressure from blasting of bedrock will be performed during April 2018. The plan so far is that measurements are to be made at three various distances from the explosion. This is to get an overview of a spread pattern depending on blast-size (amount of explosives, single/sequential blast). If possible, dependent on time and weather conditions, the NCA would like to test mitigating measures such as a "bubble curtain" and covering/capping of the blast site with gravel. There will be live cod stored in the fjord (**Figure *3***, marked with red star). Fish reactions to the blasts will also be observed and commented on in the final report for this study.

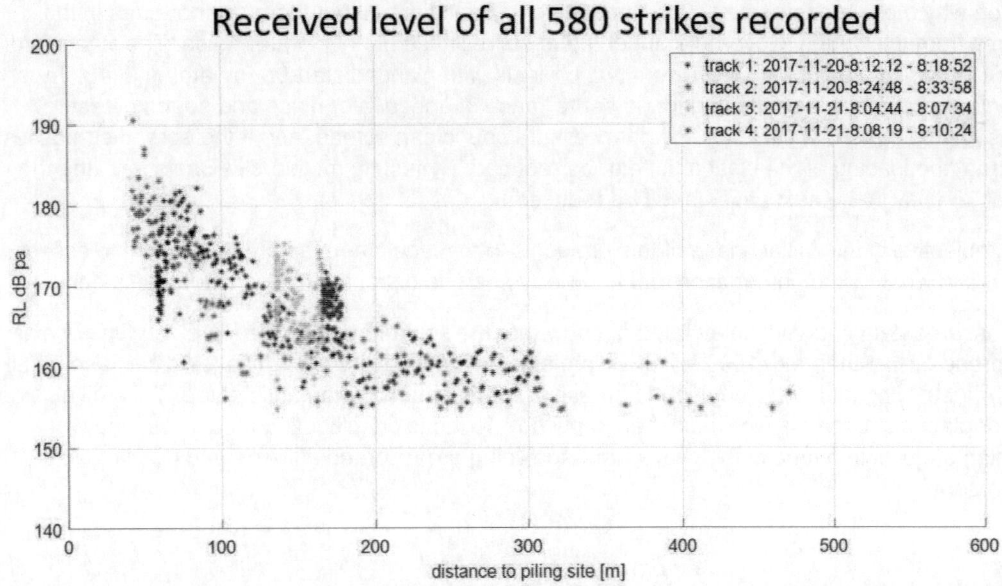

Figure 2 Track of recorded piling noise (J. Koblitz, unpublished).

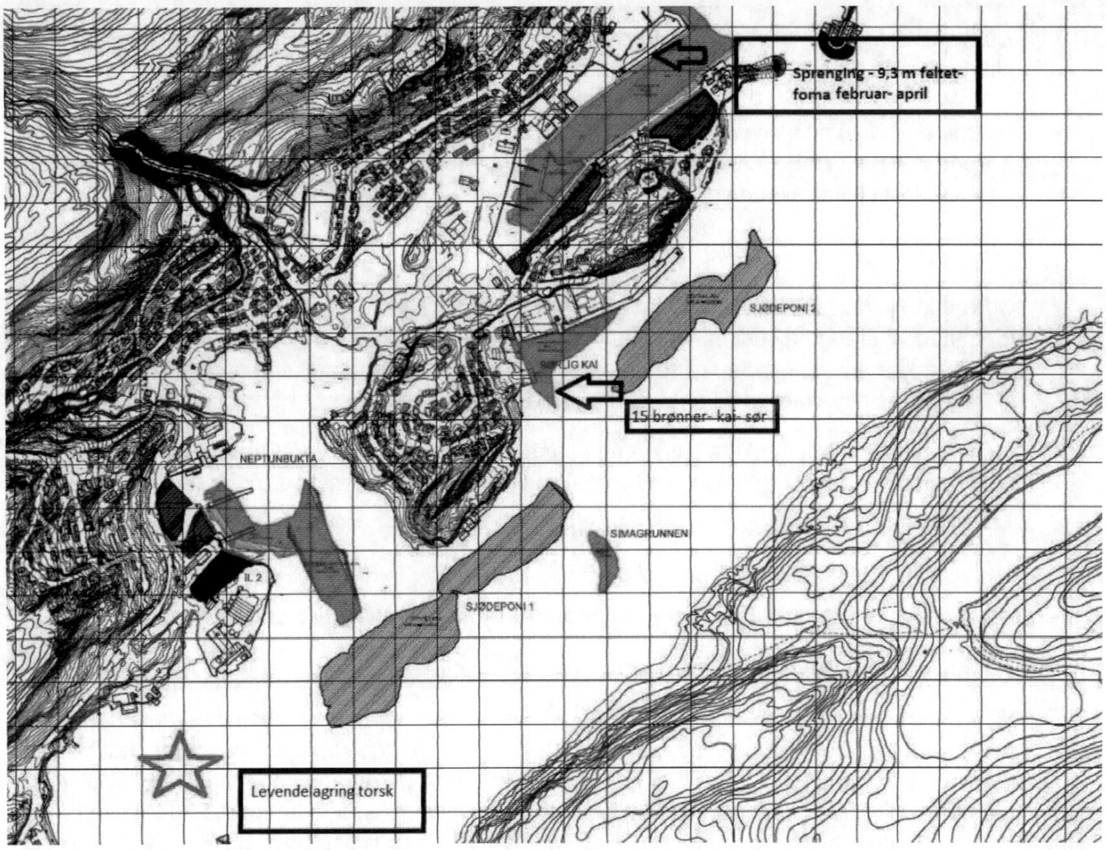

Figure 3. Construction plan for Baatsfjord. Red areas are for increased depth, grey indicates dumping site. Red star marks area for live storing of cod (Map: NCA).

The reason why these measurements are conducted in Baatsfjord is that there are no terms in the work permit from the County Governor connected to surveillance of underwater noise. The Baatsfjord-project will be a pilot project before the next two projects with planned start-up for later in 2018. These two projects do have terms connected to the surveillance of underwater noise and sound pressure. The terms have a threshold value for how far pressure sound can spread, and if the spread surpasses a given area, the loading of the blast must e.g. be reduced. By testing methods in Baatsfjord, the risk for causing a delay in the next projects will be reduced.

The main objective in this initial phase of this project is to test equipment and methods and to assess how much and what type of information and knowledge can be extracted for the work that is done.

If quality assured data and verifiable methods show that the sound pressure from a given type of blast will not spread further than e.g. 1000 m, future projects will be easier to plan. The tests can also result in the NCA identifying that noise and sound pressure spreads further than anticipated. This will also be positive for planning, because we would then know how to focus on mitigating measures, and when in doubt – plan projects to a time of the year where spawning areas, breeding birds and pods of whale are not present.

THE NCA ENVIRONMENTAL STANDARD FOR PROJECTS

All NCA projects go through a checklist in the initial phase. The need for a number of environmental surveys are assessed. Mapping and risk assessment based on factual knowledge are crucial for project planning and execution. These are typically:

- *Environmental geological and geotechnical surveys:* Site conditions must be known, both for planning of fairway projects and establishing new infrastructure.

- *Mapping of biological diversity and habitat types:* It is important that when planning projects, it is done in a way that e.g. infrastructure such as a wave breaker do not fragment habitat or damage local flora and fauna.

- *Mapping of currents and tidal patterns*

  Currents: Fjords can have a very local current pattern. This is important to map so that when, for instance, dredging sediments it can be done in a controlled way that does not result in unwanted harmful spreading of sediments. This applies both to polluted sediments as well as clean sediments, which shall not harm marine flora or fauna.

  Tide: Due to the length of the Norwegian coast, the tide will also vary greatly. The difference between the highest and lowest astronomical tide is from 0,72 meters in Oslo to 3,95 meters in Vadsoe.

Other measures taken to maintain the environmental focus in NCA projects are:

The Environmental compliance plan (EC-plan)

During 2016 a thorough revision was made of the NCAs EC-plan template. While the implementation of this plan in ongoing, the NCA has received good feedback from the supervisory bodies that have inspected our construction sites. Even though the sites will have different contractors, the EC-plan is to be implemented for all. The plan will be adapted to local conditions, and an environmental risk analysis is to be performed for each specific site. The plan is a discussion point on weekly on-site construction meetings, and is to be revised if needed. The EC-plan is a part of the NCAs environmental focus, and includes everything from accidental spills, not letting machinery run idle to the sorting of trash.

Environmental requirements in procurement is standard from 2018. It is an increased focus on not just consider price when awarding a contract, the solution on how the contractor plans to safeguard the environment will also be assessed. All state transport agencies are implementing this procedure at the same time. This is also in line with demands given from the acting government. The reduction of greenhouse gas (GHG) emissions comes under the same "umbrella", and steps are made to cut GHG-emissions with 50 % by 2030.

Another environmental issue is plastic debris from charging tubes in the sea floor, the lack of alternatives to nonel initiation systems and the plastic waste from detonating cords (**Picture *3***). The NCA does not want to be a contributor to the already enormous plastic-waste issue in the world's oceans. Therefore, we continuously work on reducing waste from our activities. One way of doing this is to have a contractual term for collecting plastic used in the project. If e.g. 500 kg of detonating cords and other plastic is used while blasting, the contractor must collect as much as possible after a blast. In addition a "positive plastic account" is set for the project, meaning that with using 500 kg plastic, 1000 kg must be collected (including detonating cords) and plastic found along nearby coastline. The contractors have so far not done the majority of the shore-cleaning themselves, but payed local volunteer organizations or school classes to do this. By this, the local community will get a better fairway, cleaned shoreline and income for a good cause or extra funding for a class excursion.

Picture 3. Detonating cords and explosives, a source for plastic waste in the sea (Photo: NCA).

CONCLUSION

Through the mentioned case-studies, the NCA aims to close some of the gap regarding the lack of factual knowledge the challenges NCA projects faces. By publishing our findings and collected data we hope to contribute to further work in this field, both nationally and internationally.

The NCA will continue to develop and improve our effort for preserving the environment. By this we also fulfill the goal set for our organization:

to make our coast and waters the safest and purest in the world.

Acknowledgements: Sound measurements in Kvalsundet was performed by Dr. Jens C. Koblitz, BioAcoustic Network, for the NCA. Report and results to be published during first half of 2018.

References

(1) The Norwegian Government, www.regjeringen.no

Environmental and societal benefits of Onshore Power Supply for Inland Navigation in Flanders (Belgium)

by

Mohssine El Kahloun[1], Pieter Vandermeeren[2], Marleen Coenen[1] and Ilse Hoet[1]

Keywords: Onshore Power Supply, Ship at-berth emissions, Port areas, Air quality, Environmental impact assessment

ABSTRACT

The use of the auxiliary engines by ships when at berth causes greenhouse gas emissions, air quality emissions and noise pollution in the port areas, which are often located in or near cities. Onshore Power Supply (OPS) is an option to provide electricity to the ships from the national grid and to reduce the unwanted environmental impact of ships at berths.

In recent years, the interest in the use of OPS has strongly increased in the Flemish ports and inland waterways. One of the actions to encourage the expansion of onshore power facilities was the setup of the Flemish Shore Power Platform (www.binnenvaartservices.be) which coordinates all actions related to the use, implementation and expansion of this environmentally friendly technology for inland navigation in Flanders.

Data on electricity consumption by a specific ship from two European projects (TEN-T project "Shore Power in Flanders" and CLean INland SHipping project) has been used to quantify the benefit of reducing the emissions of NOx, SO2, PM and CO2 that would occur by using onshore power supply.

Emissions through the use of auxiliary engines (diesel-related emissions), Emissions through the use of OPS (electricity-related emissions), and Net reduced emissions through the introduction of OPS have been calculated.

The results demonstrated that OPS can significantly reduce diesel-related emissions from ships at berth. Through the introduction of OPS the emissions of NOX can be reduced by about 93%. The emissions of PM10 can be reduced by 99% and the emissions of SO2 by more than 96%. The emissions of CO2 can be reduced by more than 90%. The reduced emission of C02 in this study is high compared to other studies, this is due to the low CO2 emission factor for electricity production in Belgium that we applied in our calculations.

In 2016, the use of OPS at only two locations has generated a total societal benefit of € 53,814. The overall use of OPS for only inland navigation in 2016 in the port of Antwerp was 766 MWh, represents a societal benefit of € 108,381.

We conclude from our analysis of the evidence that providing an onshore power supply for vessels at berth can result in significant environmental and societal gains. A communication strategy should be put in place, focusing on adequately informing and thereby stimulating the use of onshore power supply. The results of this paper can be used as basis information to convince the ship-owners of the environmental and societal benefits of OPS. River cruisers have higher power and electricity demand providing a better business case for OPS for inland navigation and a better prospect for market development. Policy makers could produce more net societal gain by implementing incentives and mandates to encourage more shift toward OPS.

---

[1] Flemish Ministry of Mobility and Public Works, Belgium

[2] Antwerp Port Authority

1. INTRODUCTION

Inland waterway transport plays an important role in the transportation of goods within Europe. In the EU27, navigable waterways stretching over 43,000 kilometres connect hundreds of cities and industrial regions. Because it's connected to Europe's most important markets, Flanders is the starting point for major freight transport via inland waterways. The river and canal network of Flanders is one of the most dense in the world, extending over 1,357 kilometres.

Inland waterway transport is an emission category that may be relevant for air quality in the vicinity of busy navigation routes or ports. Emissions from inland shipping are usually reported under the source sector non-road transport.
Shore power is the notion which indicates that a ship is connected to a shore side electrical power unit for its power supply on board. Ships at berth basically use their own generators to produce electricity. Running diesel engines in the port however, is unnecessary stressful for the environment. It produces mainly emissions of CO2, NOx, SOx and fine PM10 particles.

During the last years, the interest in the use of Onshore Power Supply (OPS) has strongly increased in the Flemish ports and inland waterways. The continuous expansion of OPS facilities also contributes to the implementation of the Flemish 3E Inland Navigation Covenant of 2009 and the 3E Inland Navigation Plan, aiming amongst others at a significant reduction of CO, NOx, fine particles and CO2. The Air Quality Plan approved on March 30th, 2012 by the Flemish Government containing measures to achieve the proposed NO2 concentrations in 2015, also foresees actions to encourage the use of shore power facility. Meanwhile, the measures for inland navigation (shipping) of the Air Quality Plan were adopted by the Government of Flanders on 30 March 2011 and must therefore effectively be implemented.

Figure 1. Overview of OPS network in Flanders.

The first action to encourage the expansion of shore power facility has been the setup of the Flemish Shore Power Platform (FSPP) (www.walstroomplatform.be) which coordinates all actions related to the use, implementation and expansion of OPS for inland navigation in Flanders. The Flemish shore power platform is involving the Flemish inland waterways managers, port managers, shippers organisations, ports and water policy officers, and stakeholders.

Several OPS-projects at local scale have been conducted by the partners of the FSPP. One of those projects is the TEN-T project (Shore power in Flanders_2012-BE-92063-S). The project's overall objective was to establish an OPS network (Figure 1), including the development of a management and payment system, for inland navigation in Flanders to contribute to its development as an environmentally friendly alternative to road transport. This management system provides the OPS managers with real time data on electricity consumption by a specific ship in a specific location. This data has been used, in the framework of the CLINSH - CLean INland SHipping project, to assess the environmental and societal benefit of using OPS in inland navigation.

This paper aims to demonstrate that providing an onshore power supply for vessels at berth can result in significant environmental and societal benefit by reducing emission of NOx, SO2, PM and CO2.

2. ESTIMATION OF EMISSION REDUCTION

2.1 Use of data and technical characteristics of the OPS system in Port of Antwerp.

Data on electricity consumption by a specific ship in 2016 at quay K75 and at quay K15 in the Port of Antwerp has been used to estimate emission reductions (NOx, SO2, PM and CO2) by using onshore power electricity.

At K75, nine new onshore power supply systems were commissioned in September 2014 (Figure 2). Seven of these consist of 4 connection points each (1 x 63 A; 2 x 32 A; 1 x universal socket of 230 V). The other two consist of 3 connection points (2 x 63 A; 1 x 125 A), which are dedicated for liquid bulk tankers. The onshore power supply systems at K15 were commissioned in January 2016 and are dedicated for use by river cruises. At this location, four systems consist of 2 connection points each (1 x 400 A; 1 x 125 A).

Figure 2. Shore power installations (low voltage) at quay 75 in Port of Antwerp

In 2016, a total of 203 cargo ships, 183 river cruises, 15 tank ship, and 50 towboat have used the OPS installations at quay K75 and at quay K15. The total electric power consumption of the shore power boxes at K75 and K15 combined, was 19762 kWh for Cargo ship, 2273 kWh for Tank ship, 7066 kWh for Towboat, and 349874 kWh for river cruises.

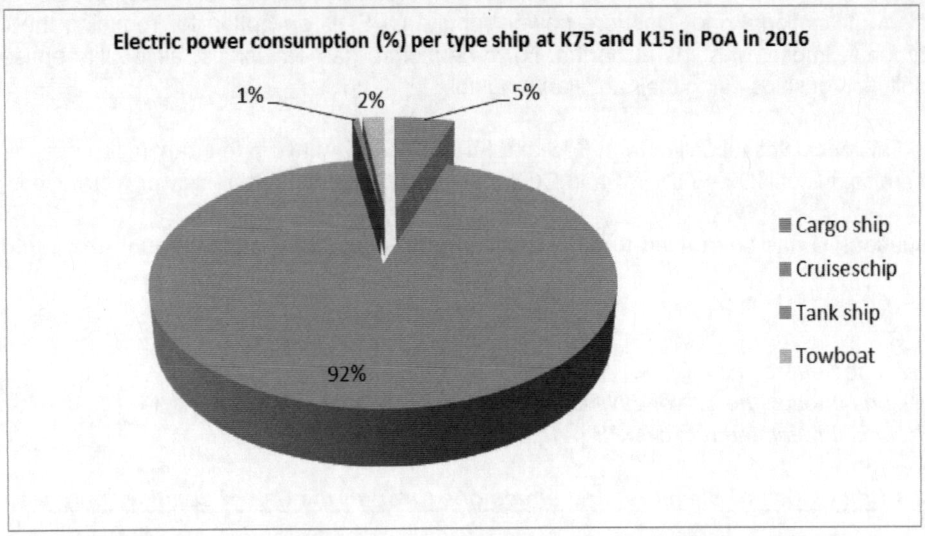

Figure 3. Electric power consumption (%) per type ship at K75 and K15 in Port of Antwerp in 2016.

The electric power consumption of the river cruises represent more than 92% of the total electrical power consumption (Figure 3). However the cargo ships spent longer time at berth (61 %) than river cruises (22%) (Figure 4).

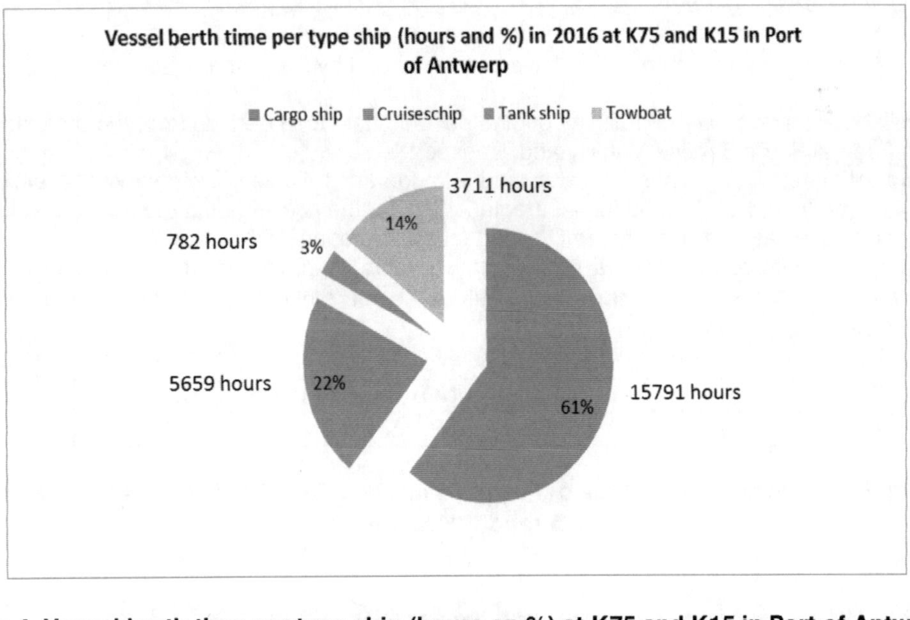

Figure 4. Vessel berth time per type ship (hours en %) at K75 and K15 in Port of Antwerp in 2016.

River cruises have higher power and electricity demand and thus provide a better business case for OPS and better prospects for market development.

2.2 Assessment to quantify the benefit of reducing NOx, SO2, PM and CO2 emissions.

One measure to reduce emissions from Auxiliary Engines while at berth is to provide electricity to the ships from the national grid. Onshore power supply (OPS) is an option for reducing the unwanted environmental impact of ships at berths, i.e. greenhouse gas emissions, air quality emissions and noise pollution of ships using their auxiliary engines.

We used historical vessel call data at K75 and K15 in Port of Antwerp to quantify the benefit of reducing the emissions of NOx, SO2, PM and CO2 that would occur if onshore power were used.

The equation (1) has been used to calculate the net reduced emissions through the introduction of OPS :

$$A = B - C \qquad (1)$$

A: Net reduced emissions (kg)
B: Emissions through the use of auxiliary engines (kg)
C: Emissions through the use of OPS (kg)

2.2.1 Calculation of diesel-related emissions through the use of auxiliary engines (B)

The amount of fuel used by ships during berth at the quay is a measure of the emissions. The used amount of fuel is the product of the number of ships, length, power output and specific fuel use to a certain amount of energy. The equation (2) has been used for calculating emissions from inland shipping (Denier van der Gon, & H., Hulskotte, J. 2010) :

B (kg) = Number of ships x Time at berth (h) x Power (kW) x Specific fuel consumption (kg fuel/kWh) x Emission factor (kg/kg fuel). (2)

Number and vessel berth time per ship have been provided by the management system of Port of Antwerp.
The available data on power of auxiliary engine per ship has been derived from the database of the Belgian Federal Public Service Mobility and Transport.
Accurate estimates of emissions from ships at berth demand reliable knowledge of the fuel consumption while at berth and associated fuel characteristics. For the missing data the average value of 100 kW has been used according to the TNO-report (Hulskotte et al 2008).
For the specific fuel consumption a representative value of 200 grams per produced kWh gas oil (Hulskotte et al 2008) has been used. Table 1 gives the different emission factors (g/kWh) per type of fuel used for the CO2, NOx, PM10 and SO2.

	CO2	NOx	PM10	SO2
Gas-oil	3160	50	2	2

Table 1. Emission factors per type of fuel used for the CO2, NOx, PM10 and SO2 as reported in the TNO-report (Hulskotte et al 2008).

2.2.2 Calculation of emissions through the use OPS (C)

The local emissions avoided by the introduction of OPS lead to additional emissions in other locations resulting from the generation of the electricity. The equation (3) has been used for the calculation of emissions from the use of OPS :

C (kg) = consumption OPS-electricity (kWh) x emission factors for electricity production in Flanders/Belgium. *(3)*

The emission factors that apply to the current electricity production in Belgium have been used (table 2). Those factors are relatively low in Belgium (as example. 285 g CO2/kWh) compared to the average emissions factor in Europe (402 g CO2/kWh) because of the use of the nuclear power plants and the use of renewable electricity production.

	CO2	NOx	PM10	SO2
Emission factor	285	0.325	0.005	0.06

Table 2. Emission factors (g/kWh) for electricity production in Belgium (Milieurapport 2016).

2.2.3 Net reduced emissions when using OPS (A)

As mentioned previously the net reduced emissions when OPS has been used was calculated using the equation (1). In Figure 5 shows that the absolute amounts of emissions saved or reduced can be mainly found in the substance CO2 and relatively in NOx. The biggest emission reductions comes from the cargo ships and river cruises that use the most energy in absolute terms.

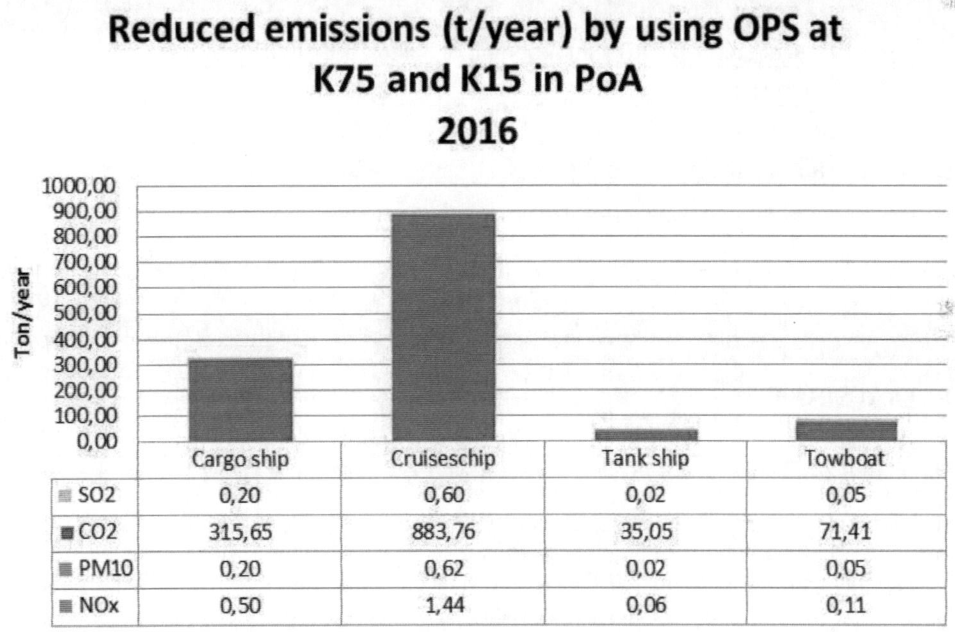

	Cargo ship	Cruiseschip	Tank ship	Towboat
SO2	0,20	0,60	0,02	0,05
CO2	315,65	883,76	35,05	71,41
PM10	0,20	0,62	0,02	0,05
NOx	0,50	1,44	0,06	0,11

Figure 5. Reduced emissions (t/year) by using OPS at K75 and K15 in Port of Antwerp.

The relative savings of emissions is more interesting because this says more about the specific benefit of the introduction of OPS. Table 3 presents the estimated percentage values of emission reduction efficiencies of OPS in 2016 at K75 and K15 in Port of Antwerp.

Type ship	NOx (%)	PM10 (%)	CO2 (%)	SO2 (%)
Cargo ship	98,74	99,95	98,25	99,42
Cruise ship	92,69	99,72	89,86	96,63

Tank ship	98,69	99,95	98,19	99,40
Towboat	98,02	99,92	97,26	99,09

Table 3. Estimated percentage of emission reduction efficiencies of OPS in 2016 at K75 and K15 in Port of Antwerp

2.3 Assessment of societal benefit of using OPS.

Using historical vessel data, we identify combinations of vessels and berth at K75 and K15 terminals that has been switched to onshore power to the largest societal benefit. The electricity consumption through the use of OPS at K75 and K15 in 2016 has been calculated.
We used the monetized societal benefit (mainly health care) when using onshore power. Taking into account the societal costs (mainly health care) of NOx, CO2, PM10 and SO2, we assumed that the use of 1 MWh onshore power has a societal benefit of 141.49 euro (AEA Technology Environment 2005). This societal benefit has been used for all combinations of vessels and berth in 2016 at K75 and K15 in 2016.

Table 4 shows the monetized health benefit by using OPS instead of burning fuels while ships are at berth. It is clear that river cruises have the most societal benefit when using OPS while at berth. This is because of the high electricity consumption by the river cruises.

Type of ship	Societal benefit of using OPS (EURO)
Cargo ship	**2,806**
Cruiseschip	**49,682**
Tank ship	**323**
Towboat	**1,003**

Table 4. Potential societal benefit (Euro) of using OPS at K75 and K15 in Port of Antwerp in 2016.

3. DISCUSSION

Proper estimation of shipping emissions is essential for an impact assessment of shipping on air quality and health in port cities and coastal regions. However it is important to stress that the objective of the current report is not to report on the methodologies to estimate emissions from inland shipping. The methodology for calculating emissions from inland shipping has been described by several reports (Hulskotte et al., 2003c; Oonk et al., 2003a; Klein et al. (2007); Hulskotte H.J & Jonkers S 2008; Hulskotte & Van der Gon, 2010 …). The objective of this paper is to demonstrate the environmental and societal effects of using OPS in inland navigation by using the methodology described in Van der Gon, H & Hulskotte, (2010). The relative savings of emissions is more interesting because this says more about the specific benefit of the introduction of OPS.
Inland shipping is an emission category that may be highly relevant for air quality in the vicinity of busy navigation routes or ports.

The results demonstrated that OPS can contribute to a largely emission reduction at local level. Figure 5 shows that the greatest reduction has been noted for the CO2 compared to the other pollutants. This is mainly due to the fact that the CO2 emission factor that apply to the current electricity production in Belgium is relatively low (285 g CO2/kWh compared to the average emissions factor in Europe 402 g CO2/kWh), because of the use of the nuclear power plants and the use of renewable electricity production in Belgium.

This is an very important issue. If renewable energy from water or wind is used, CO_2 emissions will be zero or near–zero, thus clearly giving the greatest CO_2 reductions. In Belgium, most electricity suppliers are able to supply renewably-sourced power. Implementation of OPS provides an opportunity not only to improve air quality, but also to reduce emissions of carbon dioxide, one of the main contributors to global warming. By switching from fuel oil to gas as an energy source or, better still, to sustainably green generated wind power, for example, CO_2 emissions can be curbed.

Figure 5 shows also that the greatest emission reductions comes from the cargo ships and river Cruises that use the most energy in absolute terms. The assessment suggests that onshore power may be most effective when applied at quays with a high percentage of frequently returning vessels, typically river cruises and cargo ships. The electrical power consumption of the river cruises represent more than 92% of the total electrical power consumption (Figure 3). However the cargo ships spent longer time at berth (61 %) than river cruises (22%) because of the highest number of cargo ships compared to river cruises (Figure 4). This means that the river cruises having higher power and electricity demand and less time at berth (compared to cargo ships) may provide a better business case for OPS for inland navigation and a better prospect for market development.

The relative savings of emissions in percentage is more interesting because this says more about the specific benefit of the introduction of OPS. In term of percentage of emission reductions, it seems that the emissions of PM_{10} can almost be completely avoided by 99% by the introduction of OPS. Emissions of SO_2 can almost be completely avoided by 99% for Cargo ship, Tank ship and Towboat and by 90% for Cruise ship. Emissions of CO_2 can almost be completely avoided by 98% for Cargo ship, Tank ship and Towboat and by 90% for Cruise ship. The reduction emission of C02 in this study is high compared to other studies (between 50% and 70% as reported as reported in the TNO-report by Hulskotte et al 2008, this is due to the low CO_2 emission factor for electricity production in Belgium that we applied for the calculations.

The percentage emission reduction has been estimated for a particular vessel, at berth when connected to shore power. Factors such as power consumption rate, type of auxiliary engine, type of fuel and total time at berth as described in the assessment were used to relate the overall effectiveness of onshore power. Because these factors must be evaluated for each situation, percentage of total emission reductions may slightly vary from vessel to vessel. The methodology can be used to further improve the emission estimates by using better local data (especially data on emissions factors for different auxiliary engines) when they become available.

In term of potential societal benefit, the local use of OPS in 2016 at only K75 and K15 has generated a total potential societal benefit of € 53,814. In 2016 was the overall use of shore-based power for only inland navigation in the port of Antwerp 766 MWh, representing a potential societal benefit of € 108,381.

The rates per kWh are dependent on the electricity price and the installation costs differ per kW for electricity connection. Due to the magnitude of shore power consumption, the rate charged to the inland navigation ships are close to the fares charged to households in Belgium, despite the high initial investment costs for electricity connection. Studies have shown that OPS can be beneficial for the ship-owners and port operators compared to generating electricity using fuel on-board, but ship-owners opinion are quite diverse about the cost effectiveness of OPS. Therefore a communication strategy should be put in place, focusing on adequately informing and thereby stimulating the use of onshore power supply. The results of this paper can be used as basis information to convince the ship-owners of the environmental and societal benefits of OPS.

Policy makers could also increase the net societal benefit by implementing incentives and mandates to encourage a shift toward onshore power supply. River cruises have higher power and electricity demand and thus provide a better business case for OPS for inland navigation and better prospects for market development.

4. CONCLUSION

Providing an onshore power supply for vessels at berth can result in significant environmental and societal benefit. Carbon dioxide emissions decrease substantially and emissions of Sulphur dioxide, PM10, and nitric oxide are reduced to a minimum. Through the introduction of OPS the emissions of NOX can be reduced by about 93%, emissions of PM10 can be reduced by 99%, emissions of SO2 by more than 96% and emissions of CO2 can be reduced by more than 90%.

Implementation of OPS provides an opportunity not only to improve air quality, but also to reduce emissions of CO2. By switching from to an green energy source generated wind power, for example, CO2 emissions can be curbed up to 99%. This result should be used as basis information to convince the ship-owners of the environmental and societal benefits of OPS.

Policy makers may play an important role to increase the net societal benefit by implementing incentives and mandates to encourage a shift toward onshore power supply. River cruises have higher power and electricity demand and thus provide a better business case for OPS for inland navigation and better prospects for market development.

River cruises have higher power and electricity demand providing a better business case for OPS for inland navigation and a better prospect for market development. Policy makers could produce more net societal benefit by implementing incentives and mandates to encourage more shift toward OPS.

5. REFERENCES:

AEA Technology Environment 2005. Damages per ton emission of PM2.5, NH3, SO2, NOx and VOCs from each EU25 Member State (excluding Cyprus) and surrounding seas, for Service Contract for carrying out cost-benefit analysis of air quality related issues, in particular in the clean air for Europe (CAFE) programme. AEAT/ED51014/ CAFE CBA damage costs.

Denier van der Gon., H & Hulskotte, J..2010. Methodologies for estimating shipping emissions in the Netherlands. A documentation of currently used emissions factors and related activity data. ISSN: 1875-2322.

Hulskotte J., Bolt E.W.B., Broekhuizen D., Paffen P. 2003c. Protocol voor de berekening van emissies door verbrandingsmotoren van binnenvaartschepen, Adviesdienst Verkeer en Vervoer (AVV), Rotterdam.

Hulskotte J., Jonkers S., 2008. Milieueffecten van de invoering van walstroom voor zeecruiseschepen, rivercruiseschepen en binnenvaartschepen in de haven van Amsterdam, TNO-rapport 2008-U-R0329/B/2/ TNO Bouw & Ondergrond, Utrecht, The Netherlands, 2008.

Hulskotte, J., Denier van der Gon, H. 2010. Fuel consumption and associated emissions from seagoing ships at berth derived from an on-board survey. Atmos. Environ. (1994) 44(9): 1229-1236. https://hdl.handle.net/10.1016/j.atmosenv.2009.10.018

Klein, J., A. Hoen, J. Hulskotte, N. van Duynhoven, R. Smit, A. Hensema, D. Broekhuizen. 2007. Methods for calculating the emissions of transport in the Netherlands, , Task Force Traffic and Transport of the National Emission Inventory, November 2007, CBS, Voorburg.

Milieurapport 2016. http://www.milieurapport.be/nl/feitencijfers/sectoren/energiesector/emissies-naar-lucht-door-de-energiesector/emissie-per-eenheid-geproduceerde-stroom/

Oonk, H., J. Hulskotte, R. Koch, G. Kuipers, J. van Ling. 2003a. Methodiek voor afleiding van emissiefactoren van binnenvaartschepen, TNO-MEP R2003/437, versie 2, November 2003.

River Engineering Measures for Economy, Environment and Society

Kirsten Wolfstein[1], Boris Hochfeld[1] & Henrich Röper [1]

KEYWORDS

Elbe estuary, river engineering measure, economy and environment, working with nature

ABSTRACT

The port of Hamburg, located at the Elbe estuary in Northern Germany, is challenged by a high maintenance dredging effort due to sediments which are transported by a powerful flood current from the North Sea into the upper estuary and port area. This leads to high costs for the port, but also to unfavorable environmental conditions for protected nature areas in the estuary. Therefore, the development of a profound sediment management strategy is essential to address these challenges. The main parts of the sediment management strategy, set-up by Hamburg Port Authority (HPA), consisting of several pillars, will be described. One of these pillars in order to amend the transport of sediments is the implementation of river engineering measures such as the realignment measure "Kreetsand/Spadenlander Busch" in Hamburg. Also, essential for a successful strategy is an appropriate communication strategy in order to create understanding and acceptance and support for plans and activities that might affect stakeholder and residents. This paper (1) introduces the main pillars of HPA´s sediment management strategy, (2) describes the construction and actual status of the pilot project "Kreetsand/Spadenlander Busch", a realignment measure, as an example for "Working with Nature" and (3) informs on the communication process within the estuary partnership "Forum Tideelbe" aiming at assessing and ranking further river engineering measures.

1. INTRODUCTION

The port of Hamburg is the third largest European (in TEU) and the largest German sea port. It functions as an important hub for import and export of German goods. Its location far in the Elbe estuary in Northern Germany, approximately 120 km away from the North Sea, provides excellent connections to the German hinterland. Also, more than 30% of the goods stay in the region. However, this location comes with a high maintenance effort due to high amounts of marine sediments which are transported upstream from the North Sea into the upper estuary and port area by a powerful flood current – and which the weaker ebb current is only partially able to transport back downstream, back to the sea (Figure 1). This phenomenon is called ´tidal pumping´ and is common in many estuaries with strong tidal influence. At the Elbe, it is intensified by the fact that the morphology of the estuary has significantly changed over the past centuries – due to natural effects and man-made modifications such as land reclamation, deepening of the fairway and cut-off of tributaries for flood protection.

The port does not only receive marine sediments, but also fine sediments, suspended matter respectively, from the river basin upstream. Natural erosion leads to a suspended matter input of roughly 600.000 tons per year. Transportation and mixing processes (with the marine sediments) depend strongly on the headwater discharge. In case of high discharge volumes, larger amounts of sediments are moved towards the sea, whereas during times of low discharge intensified sedimentation processes take place in the upper estuary and the port area.

1 Hamburg Port Authority, Neuer Wandrahm 4, 20457 Hamburg, Germany.
kirsten.wolfstein@hpa.hamburg.de

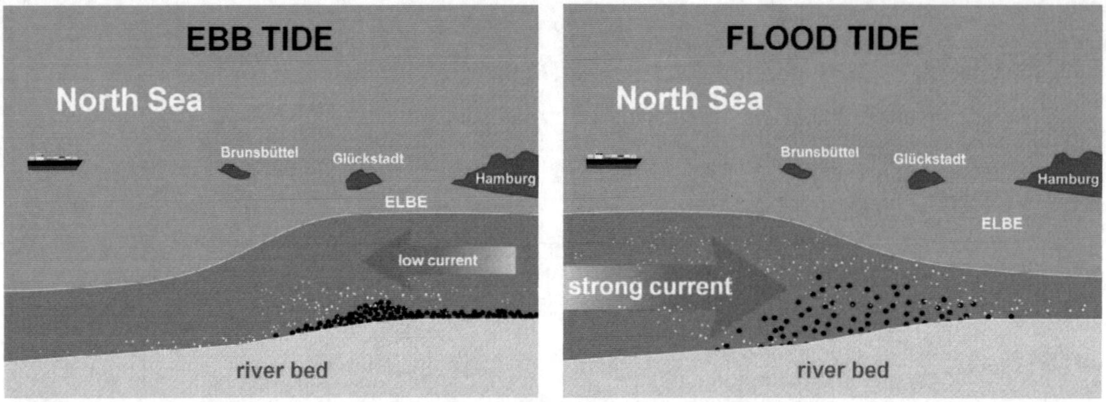

Figure 1. The tidal-pumping effect; low current during ebb tide (left); strong current during flood tide (right)

The ´tidal pumping´ effect resulted in increased sedimentation rates, especially during the last four years with very low headwater discharges, that eventually lead to unfavorable consequences for economy and ecology, i.e. high maintenance costs for the port and declining environmental conditions for protected natural habitats, especially the shallow water areas, in the estuary.

One of the main tasks for the Hamburg Port Authority (HPA) is to ensure accessibility by providing sufficient water depths for easy and secure navigation in the shipping channel within the area of the City of Hamburg and harbor basins. This results in continuous and increased dredging and disposal necessities. Since the estuary not only functions as the seaward access to the port, but at the same time is a valuable nature area that is protected by national and European legislation (EU 1992, EU 2000), HPA operates in the middle of a complex area of conflicting interests such as (1) economic and infrastructural demands of the clients of the port, (2) ecological requirements of protected species and habitats, and (3) societal interests of various stakeholders along the estuary. Therefore, HPA developed an integrative sediment management strategy which aims at addressing the described challenges, and which finally should lead to a decrease in the amount of sediments that have to be dredged.

Among other measures, sediment transport can be influenced by river engineering measures (Wurms 2017), which give more space to the river and increase the roughness of the riverbed dissipating tidal energy and reducing tidal currents. At the same time, these measures can provide benefits for nature and society. Therefore, some years ago, HPA started with the planning and construction of the pilot project "Kreetsand/Spadenlander Busch", as a realignment measure. As this measure is meant to be only the beginning of more river engineering measures in the future, an estuary partnership has been founded with representatives of all relevant stakeholders along the Elbe estuary in order to choose, examine and propose further measures to the responsible authorities.

2. OBJECTIVE

This paper introduces the main pillars of HPA´s sediment management strategy, describes the construction and actual status of the pilot project "Kreetsand/Spadenlander Busch" (as a part of one of the three pillars) and informs on the communication process within the estuary partnership "Forum Tideelbe".

Already in 2008 HPA and the national Waterways and Shipping Administration (WSV) who both are responsible of the maintenance of the tidal river Elbe set up a sediment management and river engineering concept (HPA & WSV 2008) to address the challenge of high amounts of sediments that had to be dredged and disposed. Recently HPA advanced that strategy for her area of responsibility. The current strategy consists of the three pillars "Remediation", "Maintenance" and "River Engineering", but it also includes close cooperation with relevant stakeholders and seeking for (scientific and technical) innovation (Figure 2).

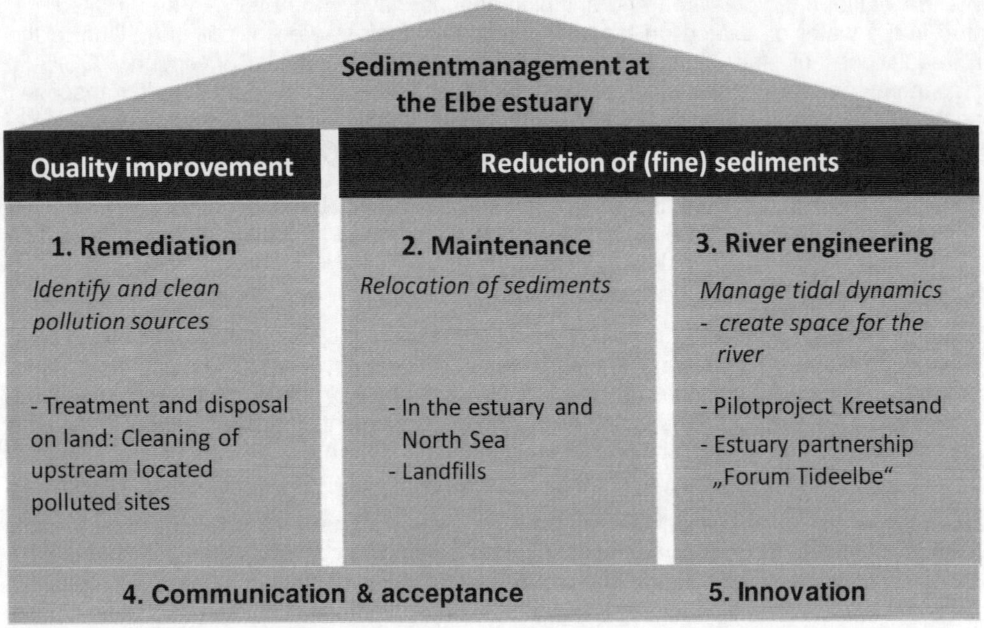

Figure 2: Sediment management strategy of HPA

The pillar "Remediation" is related to the fact that after the reunification of the two German states in 1989 water quality has greatly improved. But sediments are still contaminated with heavy metals and organic pollutants that originate from mining operations, industry and wastewater discharge and contaminated areas in the middle and upper river Elbe and its tributaries of the international catchment area (Figure 3). Depending on the amount of headwater discharge, these pollutants, bound to suspended matter, are transported to the estuary where they settle in areas of low turbulence such as the harbor basins or the shallow side areas – or eventually end up in the North Sea.

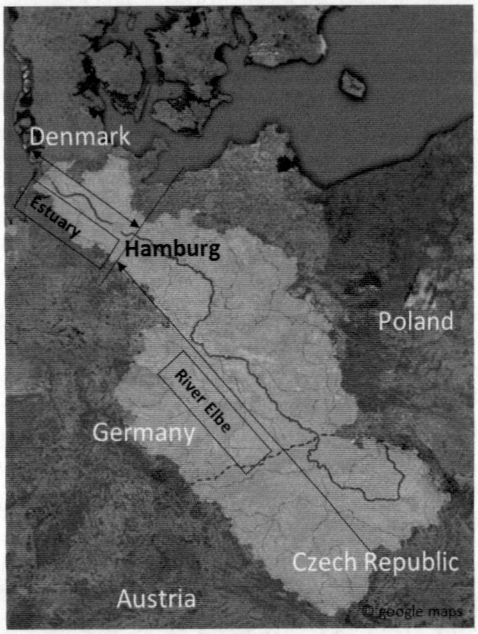

Figure 3: Map of the river Elbe and its catchment (adapted from Google maps)

The sediments that settle in the port area and shipping channel have to be dredged and treated, i.e. either relocated in the water or treated on land, depending on their degree of pollution. Due to the international regulations of the London Convention, the Oslo-Paris Convention (OSPAR, https://www.ospar.org/) and the Helsinki Convention (HELCOM, www.helcom.fi/) that were transposed into a German directive for dredged material management in estuarine and coastal areas under federal administration, sediments having a certain degree of pollution are not to be relocated in the Elbe estuary or the North Sea. But alternative treatment and placement on land leads to high costs for HPA and is not suitable for vast quantities. Therefore, the City of Hamburg and HPA took responsibility and provided 11 Million € for the project ELSA (Elbe Sediment Pollution Clean Up) which aims at the identification and remediation of polluted sediments at the sources in the upper catchment area in order to achieve, that fever contaminants reach the estuary and the port area.

The pillar „Maintenance" aims at an adaptive and flexible dredging and disposal of the sediments. Clean sediments should be relocated to different locations within the estuary or the adjacent North Sea, depending on the hydrological situation and their characteristics. Only significantly polluted sediments are to be removed and disposed of on land, using HPA's special treatment plant METHA (METHA is the acronym of **ME**chanical **T**reatment of **HA**rbor sediments).

With the pillar "River Engineering Measures" the unfavorable hydromorphological conditions are addressed that sustain the intensive tidal energy and upstream sediment transport, leading to high maintenance costs but also to unfavorable conditions for nature. In this context, measures such as the pilot project "Kreetsand/Spadenlander Busch" are planned and conducted to contribute to the reduction of tidal energy and the related heavy sedimentation and the improvement of environmental conditions for protected species and habitats. Since the Elbe estuary is an area of high economic and ecological importance, many different uses and interests have to be distinguished. Hence a close cooperation with relevant stakeholder groups is mandatory to create acceptance and avoid delays when sediment management and river engineering measures are planned and implemented. This considered, the City of Hamburg and HPA founded the estuary partnership "Forum Tideelbe" in 2016 to account for all the stakeholder's interests when suitable locations for further measures will be investigated.

Innovative technical ideas for maintenance and sediment treatment as well as the increase of scientific knowledge to understand the estuarine system and processes are also crucial. Therefore, HPA cooperates with national and international scientists and institutions and participates in (inter)national bodies and organizations.

Finally, an appropriate communication strategy to create understanding and acceptance for HPA's activities that can affect stakeholder and residents, nowadays is essential.

3. PILOT PROJECT "KREETSAND/SPADENLANDER BUSCH"

Set-up

For a couple of years, it is very well documented, especially in the US, Netherlands, Belgium, UK and other countries (Scott et al. 2011, www.tide-projcect.eu, Saathoff et al. 2013, PIANC 2018) that river engineering measures such as realignment measures are suitable to affect respectively improve hydromorphological conditions of water bodies and hence the sensitive ecological conditions of estuaries. Often, they also provide benefits for society, the so-called "Ecosystem Services" (flood protection, storage of pollutants, nutrients and CO_2, erosion and sedimentation regulation, recreation opportunities, etc., see also MA 2005). When conducting river engineering measures, concepts such as "Working with Nature" (PIANC 2018) became increasingly popular. The objectives of "Working with Nature" are consistent with those of the "Building with Nature" initiative of the EcoShape Foundation, and the "Engineering with Nature" initiative of the US Army Corps of Engineers (PIANC 2018).

River engineering measures have already been part of earlier plans. First in the sediment management concept (HPA & WSV 2008), the main focus was on reducing the tidal energy and unfavorable sedimentation. But it was recognized, that such measures could also have positive outcomes for nature and society, a so-called ´win-win´ situation. Hence the Natura 2000 integrated management plan for the Elbe estuary, issued 2012, also acknowledged river engineering measures as essential. In 2008 HPA started the planning for the pilot project "Kreetsand/Spadenlander Busch", and in 2012 the construction works began (see also PIANC 2018). The site, located in the state area of Hamburg, between the northern and southern branch of the Elbe estuary (Figure 4), was already transformed by a dike realignment carried out in 1999. However, the excavation of the high lying new foreland, a former dredged material disposal site, was not executed at that time, so that it lacked the influenced of daily tidal inundation.

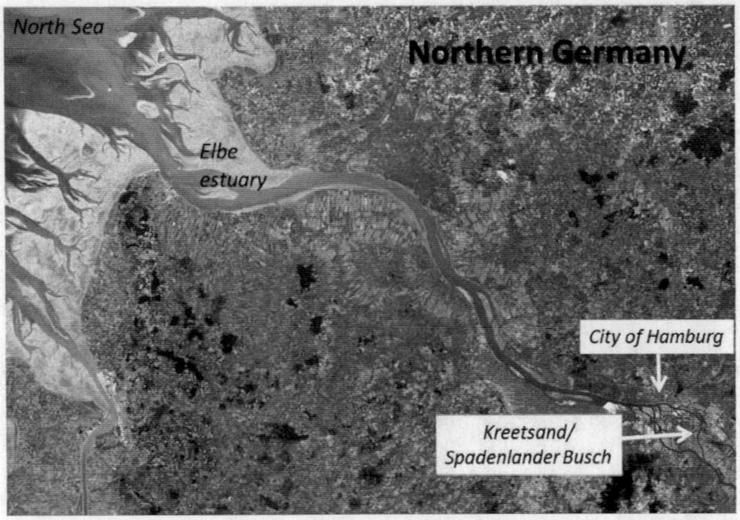

Figure 4. Location of the site (©adapted from Brockmann Consult & Waddensea Secretariat)

The measure had three objectives:

- It should contribute to reducing the tidal energy by creating approximately 1.1 mill. m^3 of additional tidal volume.
- Valuable natural habitats like shallow water areas, mudflats, reed and floodplain forests should be created.
- Parts of the area should function as recreational area for residents where they will get the opportunity to experience a tidally influenced landscape.

Approximately 47 hectares of the former flushing field were chosen to become reconnected to the estuary and influenced by the tides again (Figure 5a & b). From this area, 17 hectares were maintained as land and 30 hectares were transformed into a shallow water zone. During the planning and construction HPA already worked according the PIANC guidelines of "Working with Nature" (2018) that have been set up only recently. The approach combines social, environmental, and economic considerations into decision making, providing an integrated approach to project development and management to achieve win-win solutions for (navigational infrastructure) projects through careful consideration of natural processes, ecosystem impacts, stakeholder participation and strategies to maximize opportunities for navigation and nature. Detailed information of the planning and the subsequent construction of the measure, which received the PIANC-Award „Working with Nature" in 2014, can be found on www.pianc.org and PIANC (2018).

Figure 5a and b. The area of "Kreetsand/Spadenlander Busch" before the start of the works (©HPA) and in during the works in 2017 (© Holger Weitzel)

When works will be finished in 2020, natural processes are allowed to form the embankments inside the area by erosion and sedimentation and thus create naturally changing habitats.

However, due to the project's location in an area subject to heavy sedimentation, routine maintenance work may be necessary, prospectively every 5 years, to remove excess sediment deposits from the site. The development of the pilot project will be monitored annually for first six years, followed by an evaluation every five year. Monitoring will include topography and bathymetry measurements, habitats assessment, and the development of protected species. To better understand the impacts of the pilot project on tide surges and sedimentation the results must be analyzed in relation to monitoring results of the whole estuary.

Ecosystem Services

Meanwhile, it is widely accepted that the provision of many natural resources, which are supplied by natural ecosystems, lead to benefits for societal economy, health and survival, although often indirectly (MA, Millennium Ecosystem Assessment 2005). The capacity of ecosystems to regulate essential ecological processes through their structure, functions and biogeochemical cycles generates many Ecosystem Services (ES) (Rönnbäck et al. 2007).

Also, PIANC established in 2016 a working group that should apply Ecosystem Services Approach for Waterborne Transport Infrastructure Projects (see WG195, An Introduction to Applying Ecosystem Services for Waterborne Transport Infrastructure Projects, http://www.pianc.org/envicomactivewg.php).

In the frame of the European Interreg IVb project TIDE (Tidal River Development, http://www.tide-toolbox.eu/), the Ecosystem Services Approach was used among others by the project partners to evaluate the success of management measures within the four European estuaries Elbe, Humber,

Scheldt and Weser. This assessment offered a first screening of the potential impact on ES delivery due to the management measure. It was based on the importance of the different habitat types for the delivery of estuarine ES. Its result depended on the habitat changes caused by the measure (Figure 6). Hence, a measure will generate a positive effect on ES delivery when a habitat type with a higher importance (score) is created, as in the case of "Kreetsand/Spadenlander Busch", when adjacent land not contributing to the delivery of estuarine ES has been transformed into different estuarine habitat types, such as shallow water or marsh area which both contribute to the delivery of many different ES (see below).

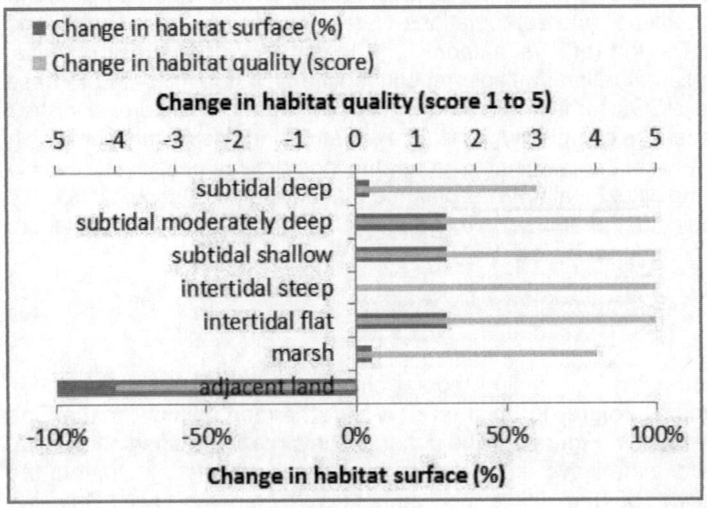

Figure 6: Ecosystem services analysis for "Kreetsand/Spadenlander Busch": Indication of habitat surface and quality change, i.e. situation before versus after measure implementation (from http://www.tide-toolbox.eu/pdf/measures/Measure_Spadenlander_Busch.pdf)

The first development target of the pilot project was the ES "Water quantity regulation: dissipation of tidal and river energy" but the ES assessment revealed that this measure generated overall a positive expected impact for many ES, mainly for:

- Biodiversity,
- Cultural services: Inspiration for culture, art and design; and Information for cognitive development,
- Regulating service: Erosion and sedimentation regulation (by water bodies).

The expected impact for the first development target "Water quantity regulation: dissipation of tidal and river energy" was only slightly positive, but the assessment revealed that the pilot measure had an even larger impact on beneficiaries for nature and society.

4. COMMUNICATION

During the last couple of years, the importance of stakeholder participation as a key to successful project design and implementation became evident, as several projects suffered a delay due to lacking acceptance and even protests from affected stakeholders and residents, for example in Stuttgart, Germany (https://en.wikipedia.org/wiki/Stuttgart_21).

However, the importance of stakeholder participation is meanwhile not only widely acknowledged but also required by environmental legislation (e.g. US National Environmental Policy Act 1970, EU 1992, EU 2000) and an important part of the "Working with Nature" approach (PIANC 2018). The development and application of a communication concept or consultation process, respectively, proved to be successful for HPA, first for the implementation of the pilot "Kreetsand/Spadenlander Busch" and secondly for the implementation of the sediment management strategy.

"Kreetsand/Spadenlander Busch"

In the "Kreetsand/Spadenlander Busch" pilot project, residents and stakeholders, including environmental organizations and local administration, were involved in the planning process right from the beginning in 2008 to inform them on HPA's plans and to obtain public input and feedback. HPA established a public information exchange and participation process for residents from the onset of the project, to introduce the project and to better understand the requirements and concerns of residents. A first information meeting took place before the planning phase commenced. The project team used public meetings to inform the local community of the project's progress and to get feedback at various intermediate phases of the work. At the end of the planning process, an overview about the planning steps, the assessment of alternative designs, and computations were provided. In the planning and approval phase, HPA worked with NGOs and regional authorities to identify and to address potential problems that could hinder project implementation. Multiple meetings with NGOs contributed to ease the approval process later, by helping NGOs remain informed of the project and by addressing their concerns. During the construction process the complexity of the tidal dynamics and estuarine functioning as well as the construction of the area itself are explained by informative posters in a publicly accessible information shed located at the site on the dike. Finally, it was planned that a "tidal park" should introduce the tidal influenced landscape and basic tidal phenomena to a wide-ranging public will be developed. The construction of the project will be completed by 2020.

Sediment management strategy

Due to Germanys constitution, federal states – including federal city states like the cities of Berlin, Bremen and Hamburg - have a certain sovereignty for their territory. For the management of the Elbe estuary this results in manifold responsibilities (Figure 7). The national Ministry of Transport and Digital Infrastructure is the owner of the federal waterways like the Elbe. In general, the management of the waterways is under the responsibility of the ministry´s authority "Waterways and Shipping Administration" (WSV) - with one exception: the management of the fairway within the city limits of Hamburg was delegated decades ago to the City of Hamburg and is now carried out by HPA. The federal states of Schleswig Holstein (northern shore) and Niedersachsen (southern shore) are responsible for the shores of the estuary, but also for the implementation of the European environmental legislation (EU 1992, EU 2000) within the waterbody of the Elbe estuary and the North Sea coastal water zone.

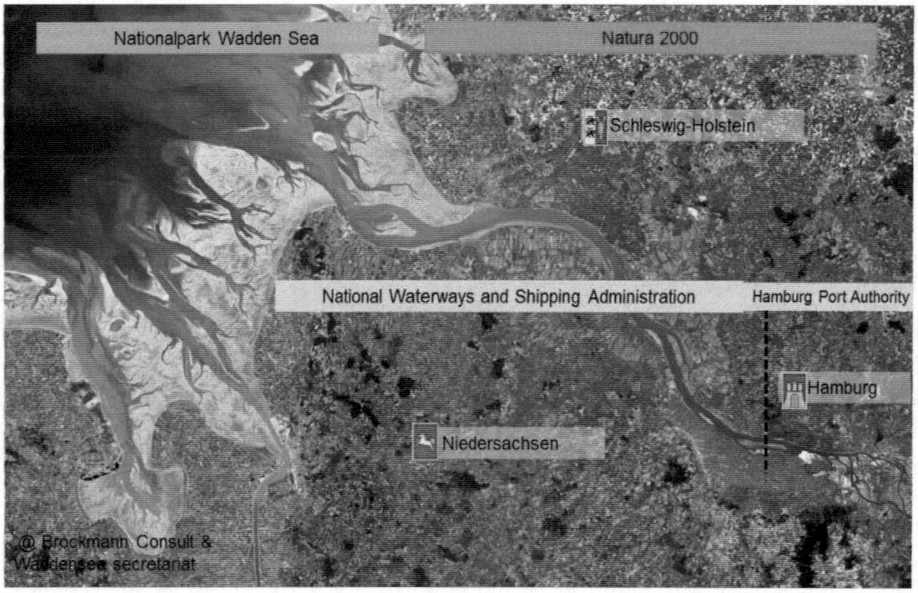

Figure 7. Location of the site (©adapted from Brockmann Consult & Waddensea Secretariat)

That is why HPA needs approval by the environmental ministry of the responsible state for the disposal of sediments in their areas. Additionally, various stakeholder groups use their influence on the administrative bodies. Therefore, HPA and WSV set up a communication process in 2014 with all the relevant stakeholders (industry, farmers, tourism, municipalities, water boards, fishermen, NGO´s, administrations of the three federal states, see figure 8) to discuss the situation and improve the sediment management of the tidal Elbe. This was crucial for the Port of Hamburg, since HPA desperately was in need of an effective disposal site outside the city state boundaries. River engineering and remediation measures with a positive effect on sedimentation quantities for the mid- and long-term perspective were also an important part of the dialogue. The character of this forum was that of a consultation body as binding decision can only be made by the responsible authorities. First in the process, a common understanding on the estuary's functioning was established and technical statements were given. Then potential locations for sediment placement and river engineering measures were discussed and finally, unanimous recommendations were given and documented in a report.

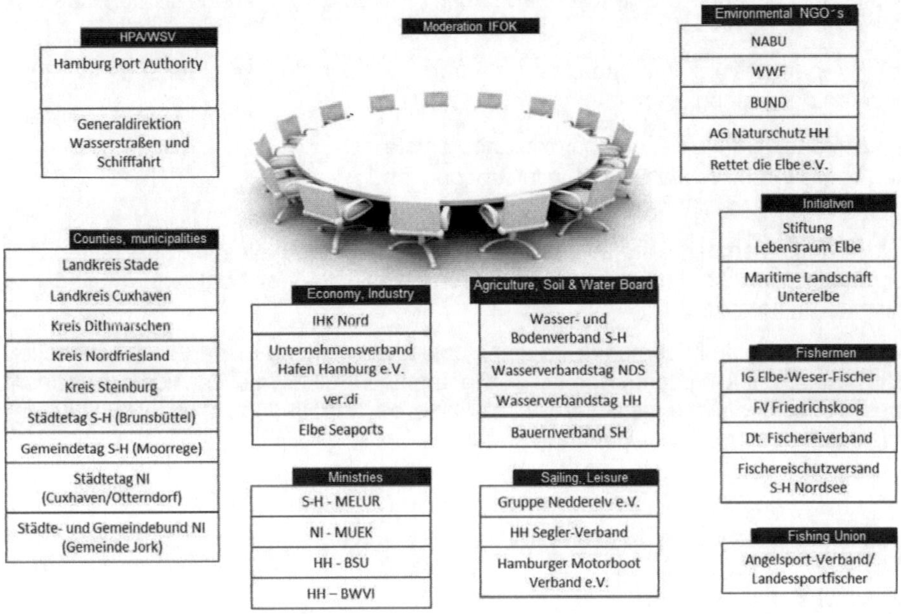

Figure 8. Members of the "Forum river engineering and sediment management tidal Elbe".

5. LESSONS LEARNED

The project "Kreetsand/Spadenlander Busch" is developing very successfully to meet the objectives of creating new aquatic and emergent habitat, and to offer opportunities for the public to interact with this important environment.

The effect of the pilot on the objective of decreasing tidal energy, however, is only marginal due to its size and its location within the estuary. To meet this target there were better locations within the estuary and more space was necessary to achieve a larger effect (Saathoff et al. 2013). However, due to the German system of federal states, this pilot project could at that time only be implemented within area of the City of Hamburg. Also, in a timely manner, there was no other suitable location available within the highly populated city area.

Nevertheless, the measure will lead to better understanding of the relationship between the estuarine system and the effects of management measures, i.e. how to improve the selection and design of a site and hence the success of the measure.

The planned monitoring and evaluation will finally assess the effectiveness of the works against the objectives of reducing the sediment transport, modifying the tidal range, and in the targets of the European Habitats Directive (EU 1992).

"Kreetsand/Spadenlander Busch" was just the beginning, more and larger river engineering measures must be planned and built. In 2015, at the end of the work intensive process of the ´Forum river engineering and sediment management tidal Elbe´ it became clear that more time was needed to investigate river engineering measures. Most certainly, finding suitable locations for these measures will be complicated by the individual interests of different federal states and various stakeholder groups.

6. FUTURE STEPS

As it became clear, that the effect of the pilot project "Kreetsand/Spadenlander Busch" on the objective „dissipation of tidal and river energy" was only marginal, common agreement rose, that further and larger measures had to be conducted to achieve significant effects. Consequently, the stakeholders involved in the "Forum river engineering and sediment management tidal Elbe" demanded to continue this process.

Therefore, an estuary partnership, the so-called "Forum Tideelbe" was founded in 2016 (http://www.forum-tideelbe.de). It should again function as

- a platform for stakeholder information and exchange of interests,
- a consultation board for the responsible administrations, and
- a trust building institution for the region of the estuary.

However, the prime task of its members is to assess and rank further river engineering measures that can lead to further stabilization of the sediment transport, dissipation of tidal energy and the establishment of valuable nature areas.

Potentially, 23 measures are eligible to contribute to these objectives. Among these, the reconnection of anabranches to the main estuary channel or the realignments of dikes are considered as most effective. All measures are located along the estuary within the area of the federal states of Schleswig-Holstein and Niedersachsen and the City of Hamburg (Figure 9).

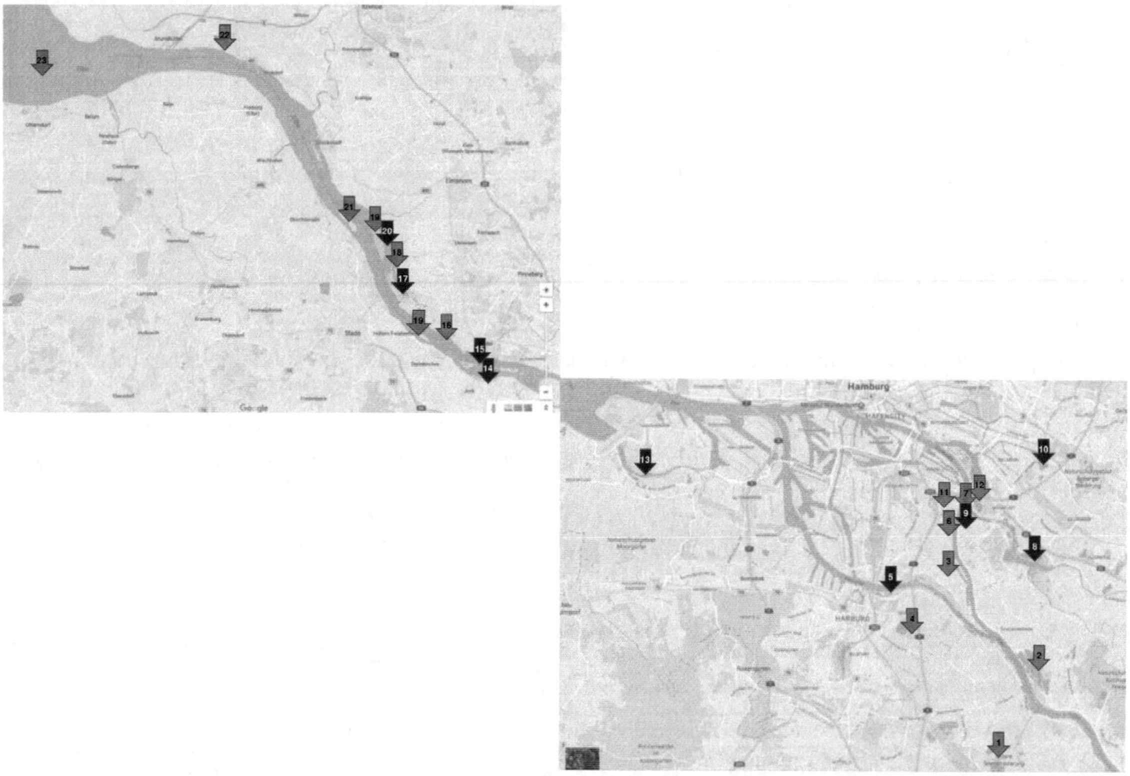

Figure 9. Locations for potential river engineering measures along the Elbe estuary (left side = mouth of the estuary, right side = upper estuary, City of Hamburg, © google maps).

The "Forum Tideelbe" started with establishing a working group to develop a pre-ranking of measures on basis of expert knowledge. This has been completed by the end of 2017. Main criteria have been the contribution of the measures to (1) reducing tidal energy and the related upstream sediment transport, (2) creating ecological valuable habitats, and (3) their feasibility (interference with flood protection, availability of property, legal concerns, etc.). Six measures have been pre-selected. They now will be considered in detail by technical feasibility studies. Also, ecological investigations will be conducted in 2018 and 2019. At the end of the process in 2020, the "Forum Tideelbe" will hand over its findings and its recommendation on the implementation of further river engineering measures to the political-administrative level. The product will consist of a well-documented report on the procedure of the ranking process, the results of the feasibility studies as well as agreements and dissents of the group concerning the proposed measures. Based on these recommendations the responsible administrative bodies will take their decisions.

As of this writing, this kind of process, i.e. involving various societal groups in estuarine management, is rather unique in Germany.

7. REFERENCES

European Commission (1992). Council Directive 92/43/EEC of 21 May 1992 on the conservation of natural habitats and of wild fauna and flora. http://ec.europa.eu/environment/nature/legislation/habitatsdirective/index_en.htm

European Commission (2000). Directive 2000/60/EC of the European Parliament and of the Council of 23 October 2000 establishing a framework for community action in the field of water policy (Water Framework Directive). http://ec.europa.eu/environment/water/water-framework/index_en.html

HPA & WSV (Hamburg Port Authority &, Wasser- und Schifffahrtsverwaltung des Bundes) (2008). River Engineering and Sediment Management Concept for the Tidal Elbe. German version "Strombau- und Sedimentmanagementkonzept für die Tideelbe". https://www.hamburg-port-authority.de/de/hpa-360/

http://www.tide-toolbox.eu/. Accessed on 5 January 2018

http://www.pianc.org. Accessed on 5 January 2018

https://en.wikipedia.org/wiki/Stuttgart_21. Accessed on 5 January 2018

https://www.helcom.fi

https://www.ospar.org/

http://www.dialogforum-tideelbe.de

http://www.forum-tideelbe.de

MA (2005). Millennium Ecosystem Assessment – Ecosystems and Human Well-being. Synthèses. http://www.millenniumassessment.org/documents/document.356.aspx.pdf.

PIANC (2018). Report No. 176-2018. Guide for Applying Working with Nature to Navigation Infrastructure Projects.

Rönnbäck, P., Kautsky, N., Pih,l L., Troell, M., Söderqvist, T. & Wennhage, H. (2007). Ecosystem Goods and Services from Swedish Coastal Habitats: Identification, Valuation, and Implications of Ecosystem Shifts. Ambio 36 (7): 534-544.

Saathoff, S., Knüppel, J., Manson, S. & Boerema, A. (2013). Management measures analysis and comparison. Investigation of measures planned and implemented at the estuaries of Weser, Elbe, Humber and Scheldt. Study report in the framework of the Interreg IVB project TIDE. http://www.tide-toolbox.eu/pdf/reports/Managment_measure_analysis_comparison.pdf

Scott, C., Armstrong, S., Townsend, I., Dixon, M. & Everard, M. (2011). Lessons learned from 20 years of Managed. Realignment and Regulated Tidal Exchange in the UK. ICE Conference Paper.

US National Environmental Policy Act (1970): https://ceq.doe.gov/laws-regulations/laws.html

Wurms, S. (2017). River Rhine between Mainz and Bingen – Morphodynamic analysis of a navigational bottleneck. In: Wieprecht et al. (eds): Proceedings of the 13th International Symposium of River Sedimentation. Taylor & Francis, London. ISBN 978-1-138-02945-3.

TOWARDS AN ECOSYSTEM BASED PORT DESIGN PROCESS: LESSONS LEARNT FROM TEMA PORT, GHANA

by

Wiebe P. de Boer[1,2], Jill Slinger[3], Arno Kangeri[4], Poonam Taneja[1], Heleen Vreugdenhil[2,3] and Tiedo Vellinga[1]

Keywords: port planning and design, ecosystems, environmental impacts, mitigation hierarchy

ABSTRACT

Connecting to the trend of harmonization of port developments with nature, this paper presents a framework for the explicit inclusion of ecosystem based alternatives in the early planning and design stages of seaport developments. The framework aims to shift the focus from offsetting environmental impacts afterwards to avoidance and reduction of environmental impacts as integral part of seaport planning and design. Our framework, labeled the *ecosystem based port design hierarchy*, helps to identify ecosystem based alternatives at 4 hierarchical levels of port planning and design: (1) consideration of alternatives to port developments to meet a perceived transport capacity problem (i.e. "no-port" alternatives), (2) port site selection, (3) port layout selection and (4) selection of port structures and materials. Application of the framework to the planning and design process of Tema port expansion in Ghana in hindsight shows that ecosystem based considerations barely played a role in alternative generation and evaluation. Therefore, opportunities for environmental impact avoidance and reduction may have been missed in the decision making process. It is recognized that decision making is a multi-disciplinary and multi-stakeholder process which is not based on environmental considerations only, but requires tradeoffs with functional, operational and socio-economic requirements. Nevertheless, we believe that explicit identification and inclusion of ecosystem based alternatives as part of this decision making process, as supported by our framework, is a requirement to arrive at port developments that are (better) harmonized with nature.

1 INTRODUCTION

Due to pressing global environmental concerns there is growing attention for harmonizing human developments, such as seaports, with ecosystems (Odum and Odum, 2003; PIANC, 2011; PIANC, 2014b; De Vriend et al., 2015; Nebot et al., 2017). Connecting to this trend, this paper provides a framework for the explicit inclusion of ecosystem based alternatives in the early planning and design stages of seaport developments. Recognizing that port planning and design starts from a scoping phase including a sound understanding of the physical, ecological and socio-economic context, this paper focuses primarily on the *design of options* step. Hence, we take a perceived transport capacity problem in a country or region as the starting point for our framework. The aim of our framework is to promote ecosystem based considerations in alternative generation in the early planning and design stages of port developments in order to increase the potential for avoidance and reduction of environmental impacts. At the same time we recognize that decision making is not only based on ecosystem and biodiversity considerations, but requires tradeoffs with functional, operational and socio-economic requirements. With this study we aim to make these tradeoffs explicit in the early port planning and design stages in order to enable transparent decision making.

We distinguish different hierarchical levels in port planning and design: (1) alternatives to a port development to resolve the identified (capacity) problem (i.e. "no-port" alternatives), (2) port site selection, (3) port layout selection and (4) selection of port structures and materials (see Figure 2). Ecosystem based alternative generation can take place at all the hierarchical levels and design choices made at higher levels narrow down options at lower levels. The identification of the hierarchical levels is based on three information sources: (1) the expertise in civil engineering

[1] Delft University of Technology, Faculty of Civil Engineering & Geosciences, The Netherlands
wiebe.deboer@deltares.nl
[2] Deltares, The Netherlands
[3] Delft University of Technology, Faculty of Technology Policy & Management, The Netherlands
[4] Wageningen University & Marine Research, The Netherlands

(background of the chief author), (2) expertise from other fields (i.e. ecology, social science, economics) through open interviews with experts (see Annex A), and (3) literature review (see Table 1). This accords with Hevner et al.'s (2004) design science approach in which societally relevant and scientifically rigorous information sources co-determine the quality of the design process. The levels of port planning and design align with the steps suggested in port design handbooks (e.g. Thoresen, 2003; Ligteringen, 2012), with the exception that most of these handbooks take the choice for a port development as a starting point. According to Zheng (2015), the port location is also often taken as a starting point, implying that most of the port design exercises start from level 3 in the hierarchy. Moreover, planning and design considerations in port design handbooks are often primarily based on functional, operational and economic motivations rather than on ecosystems and biodiversity.

Taken together the different levels of port planning and design form a port design hierarchy. The focus of this paper is on the inclusion of ecosystem based considerations at each of these hierarchical levels with the aim of accounting for ecosystems and biodiversity at the highest possible level. Therefore, we label this approach as the *ecosystem based port design hierarchy*. This hierarchy aims to shift the focus from offsetting environmental impacts afterwards to avoidance and reduction of environmental impacts as an integral part of port planning and design (see Figure 2). What is feasible, at what level in the hierarchy, is situation specific.

This paper describes the *ecosystem based port design hierarchy* framework as well as the application of the framework to the case study of the Tema port expansion, located about 2 5km east of Ghana's capital Accra (see Figure 1). The paper is organized as follows: Section 2 describes the levels of the *ecosystem based port design hierarchy*. Section 3 provides the methodology for testing our framework based on the Tema case study. Consequently, the results of the case study application are described in Section 4. Finally, the conclusions are presented in Section 5.

Figure 1: Location of Tema port at the Ghanaian coastline

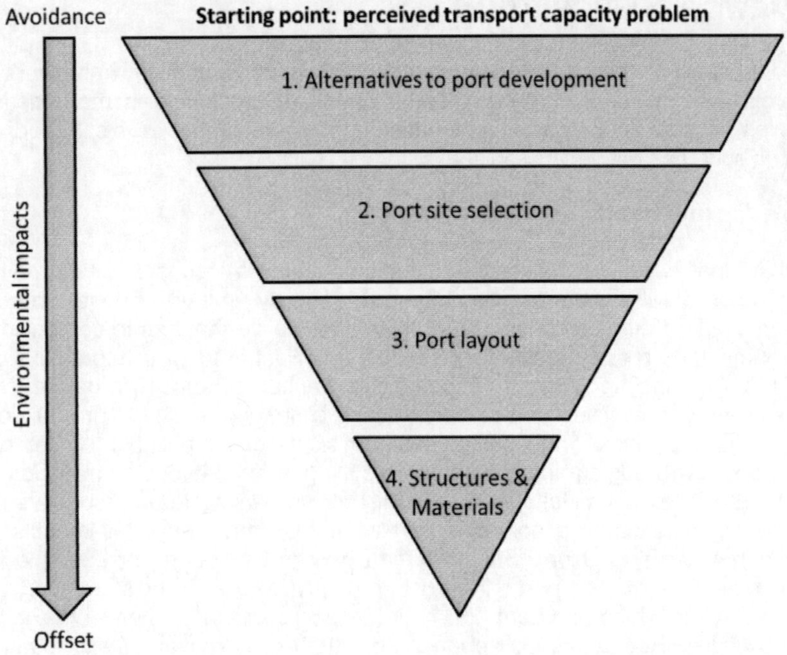

Figure 2: The *ecosystem based port design hierarchy* for avoiding/minimizing environmental impacts at the different hierarchical levels of port planning and design.

Level in port design hierarchy	Sources
1. Alternatives to port developments	Alemany 2005; Nebot et al, 2017; Robins, 2002; Xiao and Lam, 2017; Expert interviews & brainstorms
2. Port site selection	Alemany 2005; Diab et al., 2017 Nebot et al, 2017; PIANC, 2018; Schipper et al., 2015; SIGTTO, 1997 Tallis et al., 2015 Thoresen, 2003; Ligteringen, 2012; Zheng, 2015; Expert interviews & brainstorms
3. Port layout alternatives	Bakermans et al., 2014; Bruun, 1981, 1990, 1992; De Jong et al., 2012; Pachakis et al., 2017 Schipper et al., 2015 ; Expert interviews & brainstorms
4. Port structures and materials	Burt et al., 2009; De Vriend et al., 2015; Dekker et al., 2014; Mercader et al., 2017; Odum and Odum, 2003; Paalvast et al, 2012; Perkol-Finkel et al., 2017;

Table 1: Overview of information sources used for the development of the *ecosystem based port design hierarchy*

2 FRAMEWORK DESCRIPTION

As presented in Figure 2 the *ecosystem based port design hierarchy* framework consists of 4 hierarchical levels. The subsequent subsections describe these levels in more detail and discuss general examples of ecosystem based alternatives for each of the levels based on information obtained from the sources provided in Table 1.

2.1 Level 1: Alternatives to port development

The first level in the *ecosystem based port design hierarchy* considers alternatives to a port development in order to solve the perceived transport capacity problem. These "no-port" alternatives are not to be confused with the obligatory "do-nothing" alternative that has to be considered as part of a Social Environmental Impact Assessment (SEIA). As illustrated in the previous section a port development is one of the solutions to a perceived transport capacity problem, for example by expanding an existing ('brownfield') port or constructing a new ('greenfield') one. A downside of port developments is that they impact the environment in terms of air pollution, water pollution, waste disposal and, especially, dredging and civil works resulting in habitat loss/degradation (Peris-Mora et al., 2005; ESPO, 2012; Lam and Notteboom, 2014). From an ecosystem perspective it can therefore be beneficial to explore alternative solutions that have less environmental impacts to resolve the perceived transport capacity problem. Such alternatives can be sought for example in efficiency or utilization improvements of existing port infrastructure (PIANC, 2014a), redistribution or repurposing of existing port infrastructure (Xiao and Lam, 2017), increased cooperation between existing ports in the form of port networking (Robins, 2002; Nebot et al., 2017) or improvements of other transportation modes and their mutual connections (Nebot et al., 2017). Such alternatives may not avoid environmental impacts entirely, but at least minimize the need for (additional) dredging and civil works and, hence, avoid/minimize the environmental impacts associated with those activities.

2.2 Level 2: Port site selection

In many cases alternatives to a port development are limited or just not sufficient to resolve the perceived capacity problem entirely. At this stage it is assumed that a port development is inevitable. The next step in the *ecosystem based port design hierarchy* is then to select an appropriate site for the port development. From an ecosystem perspective extension and requalification of existing ('brownfield') port infrastructure is often preferred over construction of new ports (Alemany, 2005; Nebot et al., 2017). Since marine and terrestrial infrastructure are already in place for existing ports, the additional environmental impacts in these (already impacted) environments are likely to be less severe than the impacts of new ports in pristine environments. Also from an economic point of view extension or requalification of an existing port site may be beneficial compared to a new port site, due to the availability of existing infrastructure, hinterland connections, labor and supporting socio-economic infrastructure, resulting in lower overall costs for the development.

In some cases the development of a new ('greenfield') port is required, because the existing ports are located too far away from the location where additional transport capacity is needed or expansion of existing ports is impossible for physical or socio-economic reasons. In those situations port site selection can be crucial to avoid and minimize impacts on ecosystems and biodiversity (Tallis et al., 2015; Zheng, 2015; Schipper et al., 2015). From an ecosystem perspective the site selection should be such that the functioning and integrity of natural ecosystems is preserved. This means accounting for ecological requirements such as (1) habitat connectivity, (2) limited direct human interferences, (3) endogeneity, (4) species population viability, (5) opportunities for threatened species, (6) trophic web integrity, (7) opportunities for ecological succession, (8) zone integrity, (9) characteristic (in)organic cycles, (10) characteristic physical-chemical water quality and (11) system resilience (based on Slinger and Nava Guerrera, 2016). Accounting for these ecological requirements implies working with nature rather than working against or despite it (PIANC, 2011; PIANC, 2014b; De Vriend et al, 2015). Ideally, the site selection is such that the natural local conditions enable port functioning so that little human interferences are required. Such a location would be naturally sufficiently deep for navigation, allows for sufficient manoeuvring space and has sufficiently mild conditions (e.g. in terms of wind, waves and currents) to enable safe and efficient port operations (Thoresen et al., 2003; Ligteringen, 2012; PIANC, 2014a; Zheng, 2015; Diab et al., 2017). Additionally, such a site would be ecologically resilient, consisting of environments similar to those created by port infrastructure (e.g. vertical stony coastline) and neither biogeographically unique nor fulfilling a unique function for the regional ecosystem (e.g. in terms of migration routes or nursery function). Obviously, ecologically protected and sensitive areas should be excluded from the site selection process as much as possible.

2.3 Level 3: Port layout selection

After port site selection the next level in the *ecosystem based port design hierarchy* is port layout selection. The available design options for port layouts are highly dependent on the ambient natural conditions in terms of water depth, waves, wind and currents and, hence, constrained by the selected port development site. As discussed in Section 2.2, ideally site selection is such that natural conditions enable port functioning without human interferences in the natural system, but often human interferences in the form of dredging (e.g. deepening of port basin and access channels, land reclamations for quay infrastructure) and civil works (e.g. breakwaters, berths and quay walls) are required. From an ecosystem perspective these human interferences should be avoided or minimized as much as possible. Hence, port lay-outs with minimal needs for dredging and civil works are preferred from an ecosystem's point of view.

Traditionally seaport layouts consist of breakwater structures to provide shelter from ambient wave and current conditions (De Jong et al., 2012). However, in mild coastal environments, open or unsheltered port concepts have been suggested as an alternative to traditional port layouts (Bruun, 1981, 1990, 1992; De Jong et al., 2012; Bakermans et al., 2014). Such port layouts would decrease the amount of civil works required. This is not only interesting from an economic point of view, but also minimizes human interference in the ecosystem in terms habitat modification and segregation. The feasibility of open layouts depends very much on the hydro-geomorphological environment of the selected site as well as the typical vessel characteristics, port operations and available mooring and crane technology in the port (De Jong et al., 2012). Another alternative to traditional port layouts are offshore ports or terminals. Offshore ports or terminals are generally located at deeper water and, hence, require less dredging works to deepen port basins and access channels (Pachakis et al., 2017). This also holds for extensions of existing ports, where extensions in offshore direction are likely to be beneficial over alongshore port extensions in terms of required dredging works. Moreover, the coastal zone is often characterized by a higher habitat richness and biodiversity than the offshore zone due to larger heterogeneity (Gray, 1997; Ray, 1988). These richer habitats do not have to be displaced in case of an offshore extension. Although the feasibility of offshore layouts is highly situation specific, bigger ships and more sophisticated mooring systems offer more opportunities for offshore ports nowadays than in the past.

2.4 Level 4: Selection of port structures and materials

At the lowest level of the *ecosystem based port design hierarchy* the design freedom is constrained to choices with respect to the type of structures and materials to be used for the port infrastructure. Although possibilities to avoid environmental impacts at this level are generally limited, still ecosystem based design choices can be made to minimize environmental impacts, restore ecosystem functioning or enrich system complexity. As a starting point we assume at this stage that standard port infrastructure in the form of a sufficiently deep port basin with breakwaters and quay walls is required. Such infrastructure replaces natural habitats (Dugan et al., 2011). From a marine perspective such infrastructure adds a significant amount of hard substrate for marine organisms to attach to. However, the steep slopes, low structural complexity and high homogeneity of traditional port infrastructure do not provide suitable conditions for the development of diverse biological assemblages (Firth et al.,2016; Perkol-Finkel et al.,2017). Therefore, these structures are often dominated by nuisance and invasive species (Mineur et al., 2012). From a terrestrial perspective traditional port infrastructure not only homogenizes habitats, but it also excludes growth opportunities for plants and microorganisms which are essential for terrestrial ecosystems.

Over the past years advances in the field of ecological engineering (Odum and Odum, 2003; PIANC, 2011; De Vriend et al., 2015) have resulted in new concepts for multifunctional design of marine infrastructure also accounting for the ecosystem. Still this approach is rarely applied in the development of ports (Mercader et al., 2017). Examples of ecological engineering concepts for ports found in literature include breakwaters with an artificial reef function to support habitat richness and biodiversity (Burt et al., 2009), ECOncrete® to enhance the biological and ecological value of quay walls while contributing to structural integrity (Perkol-Finkel et al., 2017), artificial habitat creation to enhance the nursery function of juvenile fish by increasing habitat complexity (Mercader et al., 2017), surface complexity enhancement using novel resurfacing materials (Dekker et al., 2014) and pole and pontoon hulas (i.e. hanging ropes) to increase the productivity and biodiversity in the hard-substrate environment (Paalvast et al., 2012). If modifications to native habitats and the ecological functions they support are unavoidable due to the construction of port infrastructure, these ecological engineering options may help to create alternative habitats to sustain existing or support new ecological functions (Mercader et al., 2017).

3 CASE STUDY METHOD

In line with the design science approach of Hevner et al. (2004) we test our prototype *ecosystem based port design hierarchy* framework (or artifact) on a practical case study in order to evaluate its utility. For this we use the case study of Tema port expansion in Ghana. The aim of the application of the *ecosystem based port design hierarchy* to this case is primarily to test and illustrate our framework. With the application of our framework we (1) evaluate the inclusion of ecosystem based considerations in the planning and design process of the Tema port expansion based on available public information and (2) identify potential opportunities for ecosystem based alternatives that could have been considered at the different hierarchical levels in hindsight. Information on the case study is obtained primarily from the following sources (1) a site visit to Tema port, (2) open interviews with local stakeholders in and around the port and (3) preliminary, feasibility and environmental impact studies of Tema port expansion (JICA and GPHA, 2002; Halcrow, 2010; SAL Consult Ltd, 2015).

The port of Tema is located in the Greater Accra Region about 25 km east of Ghana's capital Accra. Presently, about 80% of Ghana's sea borne international trade, which is almost 90% of Ghana's total trade, is handled through port of Tema (Halcrow, 2010; SAL Consult Ltd, 2015). The port is mainly operated by the Ghana Ports and Harbour Authority (GPHA), except for a modern container handling facility that is operated by Meridian Port Services Ltd (MPS), a joint venture of APM Terminals International, Bollore Group and GPHA (Halcrow, 2010). The cargo consists mainly of containers, but also includes bulk, RoRo and oil & gas. Over the past years port of Tema has seen a significant increase in total cargo traffic from about 7.5 million ton in 2003 to 13.5 million ton in 2016 (GPHA, 2017). Due to the rapidly growing economy of Ghana and the potential for growth in transit trade through Ghana to the landlocked countries in the sub-region, the cargo traffic through port of Tema is expected to further increase in the future (SAL Consult Ltd, 2015). Also transhipment is a growing core activity of the port (Halcrow, 2010). To facilitate this growth a joint venture of the Ghana Port and Harbour Authority (GPHA) and Meridian Port Services Ltd (MPS) is currently expanding the port with container handling facilities (see Figure 3).

Tema port is located at the Ghanaian coastline with a gently sloping coastal profile towards the sea (SAL Consult Ltd., 2015). The port area is underlain by a system of rocks (Halcrow, 2010). The hydrodynamics are affected by the Guinea current, tide, wind and waves. The tidal current is generally less than 0.1 m/s (JICA and GPHA, 2002; SAL Consult Ltd., 2015). The alongshore currents in the nearshore are mainly driven by wind and waves with an average of about 1 m/s (SAL Consult Ltd., 2015). The waves are predominantly swell waves from the south-southwest originating from the Atlantic ocean with typical significant wave heights between 1 and 2 m (SAL Consult Ltd., 2015), resulting in an eastward directed littoral drift. Whereas the SEIA shows significant predicted sediment transport rates leading to up-drift sedimentation and down-drift erosion as a result of the port expansion (SAL Consult Ltd, 2015), indicating significant sand availability, the feasibility study of Halcrow (2010) states that suitable fill material in the vicinity of the port is scarce based on available soil borings. The latter is confirmed by stakeholder interviews during a site visit to Tema in 2017 (see Annex A). Sand in the vicinity of the port seems to be scarce and fill material for the port extension needs to be dredged from far offshore. Three lagoons are present in the immediate vicinity of the port: the Ramsar designated Sakumo II lagoon about 1 km west of the port, the almost fully silted Chemu lagoon just east of the port and the Gao lagoon about 3km east of the port (Halcrow, 2010). The Chemu lagoon serves as the major effluent drain for the industries of Tema and is heavily polluted. The Gao lagoon is also silting up and getting polluted by emerging industries in its vicinity. The Sakumo II lagoon is largely closed off from the sea after the construction of the Accra-Tema coastal road in the 1950's. Only supra tidal culverts provide a small connection to the sea. The brackish-saline lagoon consist of open lagoon, floodplain, freshwater marsh and coastal savanna habitats. It is a tourist point for sea and migratory bird watching, provides a source of livelihood for fishermen and has spiritual value for the communities living alongside the lagoon.

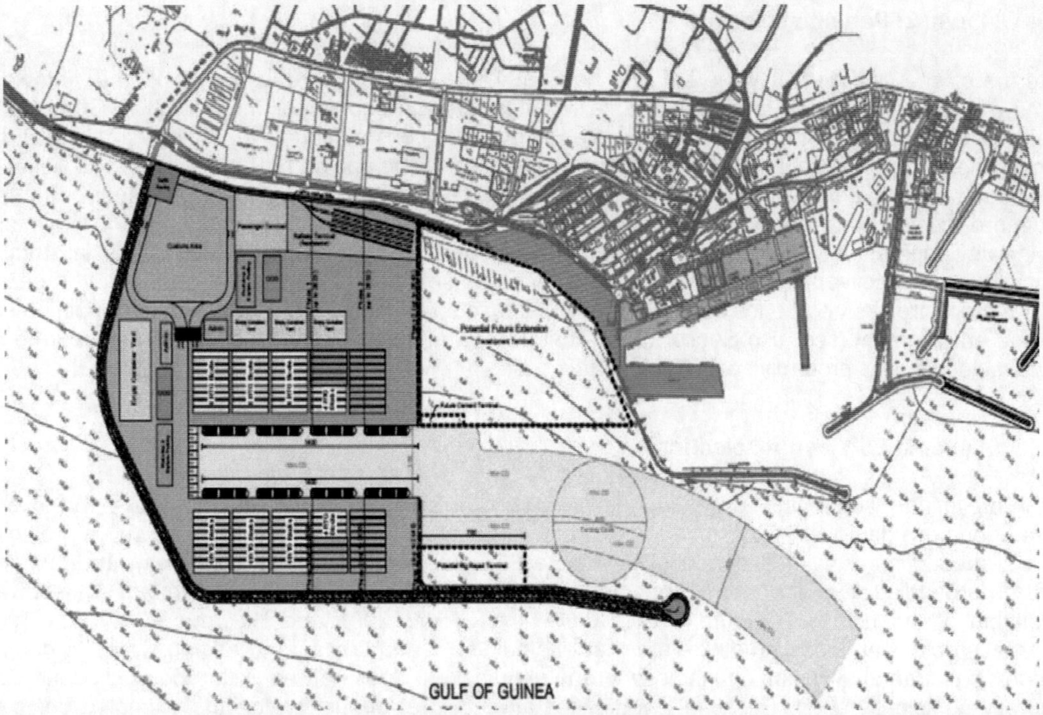

Figure 3: The planned expansion of Tema port (highlighted in full color) and the present port layout (in grayscale). Source: GPHA (2018).

4 APPLICATION OF THE ECOSYSTEM BASED PORT DESIGN HIERARCHY TO TEMA PORT EXPANSION

In this section we apply the *ecosystem based port design hierarchy* in hindsight to the case study of Tema port expansion. The aim of the application to the Tema case is both to test our framework in practice and to derive lessons learnt from the Tema case. The subsequent sections evaluate and discuss ecosystem based alternative generation at each of the levels of the port design hierarchy.

4.1 Level 1: Alternatives to port development

In the feasibility study for Tema port expansion (Halcrow, 2010) alternatives are explored in order to facilitate the expected growth in container throughput for which the capacity of the present facilities is not sufficient. International cooperation between the many West African countries with their own ports is not considered a realistic alternative, because of the lack of good coastal roads between adjacent ports as well as issues with customs and national pride (Halcrow, 2010). Therefore, the focus of the feasibility studies is on improvement of the existing port infrastructure in Ghana. In the feasibility study two noticeably different scenarios are considered: (1) a *gateway port* serving mainly Ghana and transit traffic with limited transhipment and (2) a *regional hub port* serving the above plus neighbouring countries via transhipment. The latter scenario involves much larger ships and a doubling of traffic volume in Tema compared to the former (Halcrow, 2010). In the *gateway port* scenario several alternatives are explored to upgrade the existing port infrastructure within the present breakwaters to expand the container handling facility. The limited dredging and construction needs of such a scenario would have been preferable from an ecosystem perspective. In the *regional hub port* scenario, ecosystem based alternatives to a port development in order to accommodate the larger vessels and traffic volumes for transhipment do not seem to have been feasible. The *regional hub port* scenario is eventually adopted for Tema port expansion.

4.2 Level 2: Port site selection

In terms of site selection only the existing ports of Tema and Takoradi were considered suitable in terms of size and facilities to accommodate for the increasing container throughput in Ghana. The choice of an existing port has several advantages over the construction of a new one in terms of the existing network, infrastructure and hinterland connections. The choice for Tema instead of Takoradi was based on the Tema's advantages in terms of more liner vessel calls, greater container berth draft and more modern container handling facilities, storage and management (Halcrow, 2010). The choice for the development of an existing port over the construction of a new one is also beneficial from an ecosystem perspective, because the additional impacts of a port expansion in an already impacted environment are likely to be less severe than the impacts resulting from a new port construction in a pristine environment (see also Section 2.2). Therefore, a more ecosystem based alternative in terms of site selection was probably not feasible in this case.

4.3 Level 3: Port layout selection

In the preliminary and feasibility studies for the expansion of Tema port three alternative port layouts have been considered: (1) expansion to the west with container berths parallel to the shore protected by a L-shaped breakwater attached to the shore, (2) similar to alternative 1 but with the container berths perpendicular to the shore protected by a detached offshore breakwater and (3) seaward extension of the port (see Figure 4 from JICA and GPHA, 2002 and Halcrow, 2010). The main differences between the alternatives are that alternatives 1 and 2 require dredging works to deepen the port basin and navigation channels whereas alternative 3 does not require dredging, because it is naturally sufficiently deep. Because layout alternative 3 is at deeper water, the expected costs for breakwater construction are higher than for the other alternatives. The fill for the land reclamation in alternatives 1 and 2 can partly be obtained from the dredging works, but the dredged volumes are considered insufficient for the entire fill. For alternative 3 the entire fill for the land reclamation needs to be obtained from sources on land or further offshore.

Alternative 1 was recommended by JICA and GPHA (2002) as the most desirable alternative based on the criteria and scores presented in Table 2. However, Halcrow (2010) argues that the comparison is unfair to alternative 3, because (a) the increased capacity in terms of ship size of alternative 3 due to the larger depths is not reflected in the cost index in Table 2, (b) it is not likely that there is any difference in the quality of the berths and in terms of calmness of the water and (c) there is no substantial difference between the schemes in terms of future development. Instead, Halcrow (2010) recommends alternative 3, because it leaves the area seaward of the existing container yard for future development and involves less dredging. The latter makes the investment estimate much more reliable, because the availability of sediment sources offshore of Tema is uncertain.

Environmental considerations are not explicitly mentioned in the comparison of alternatives. All alternatives get equal scores on the criterion “Harmonization with environment” whereas at first sight the environmental impacts of alternative 3 seem to be much less than the other alternatives. Firstly, alternative 3 involves much less dredging works than the other alternatives. Secondly, alternative 3 leaves the coastal zone west of the existing port unaffected. The coastal zone is likely to offer more habitat richness and biodiversity than the zone offshore of the existing port, because it is less homogenous and less affected by present ship traffic. Thirdly, alternative 3 leaves space for the (optional) restoration of the Sakumo II lagoon. Although the Sakumo II lagoon is not directly related to the port development, it leaves the option for ecosystem and biodiversity restoration of this Ramsar site open. Hence, from an ecosystem perspective alternative 3 would have been more desirable than the other alternatives.

Despite the above considerations and Halcrow's recommendation, the port layout that eventually has been selected for construction (see Figure 3) is similar to alternative 1, but extends even farther to the west, covering the entire coast from the present port breakwater to the outlet culvert of the Sakumo II lagoon. The motivation for this choice could not be tracked from the studied documents. In the Social and Environmental Impact Assessment (ESIA) of the Tema port extension (SAL Consult Ltd, 2015) only this selected layout has been considered and compared to the no action (or “do-nothing”) option. Hence, alternative, more ecosystem based layouts have not been explicitly weighed in terms of their environmental impacts. Instead, the SEIA focused on a comparison of alternatives in terms of structures and material for the selected layout, which is one level lower in our framework.

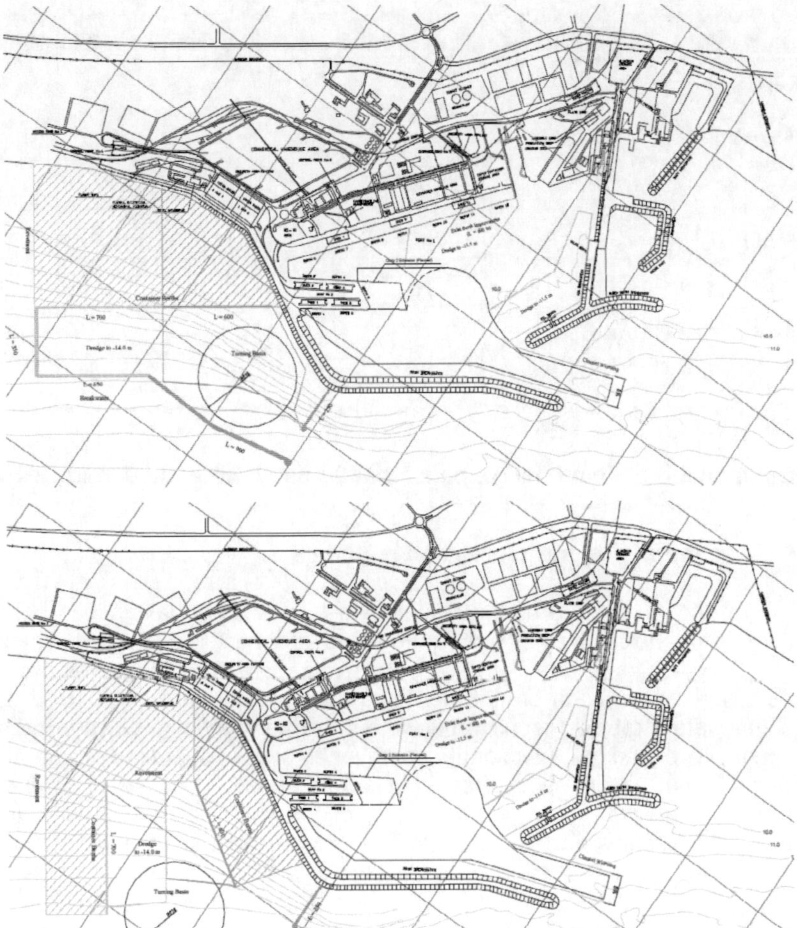

Figure 4: Alternative port layouts considered for the Tema port expansion, from top to bottom alternative 1, 2 and 3 (source: JICA and GPHA, 2002)

	Alternative - 1	Alternative – 2	Alternative - 3
Quality of Berths	***	***	**
Calmness of water	***	***	**
Navigational safety	***	***	**
Future Development	***	***	**
Disturbing existing port facility	***	***	***
Harmonization with environment	***	***	***
Cost Index	100	112	107

Note: *** Good **Fair *Poor

Table 2: Comparison of Tema port expansion layouts (source: JICA and GPHA, 2002)

4.4 Level 4: Selection of port structures and materials

At the lowest level in the *ecosystem based port design hierarchy* the layout has been fixed, which narrows down the design choices to the type of structures and materials for the port infrastructure. In the feasibility and SEIA studies for the Tema port expansion different alternatives are considered for the type of port structures in terms of breakwaters and quay walls (Halcrow, 2010; SAL Consult Ltd., 2015). For the breakwaters rubble mound breakwaters are compared to caisson breakwaters. The comparison is primarily based on functional, operational and economic considerations such as structural integrity, material needs, ease of construction and costs of construction and maintenance (SAL Consult Ltd., 2015). Ecosystem based considerations such as the potential to create more heterogeneity and niche habitats on the hard substrate as well as nature based alternatives for the "hard" engineering solutions have neither been included in the generation nor the evaluation of alternatives.

For the quay wall structures block wall, piled pier, caisson and cofferdam alternatives are considered in the SEIA (SAL Consult Ltd, 2015). Those alternatives are qualitatively evaluated in terms of buildability, durability, local experiences, availability of local equipment, adaptability to local soil conditions, wave reflection and construction costs. Again, ecosystem based considerations, such as chemical leeching from structural materials and reinforcement of existing functions are neither included in alternative generation nor the evaluation. Therefore, potential opportunities for increased habitat complexity (Mercader et al., 2017), the use of alternative materials such as ECOncrete© (Perkel-Finkel et al., 2017) and other options described in Section 2.4 may have been missed. It is not said that the inclusion of such alternatives or the incorporation of an ecological evaluation criterion would have resulted in other choices for the types of structure and materials. However, it would at least have been an explicit consideration in the design process and decision making.

5 CONCLUSIONS

Based on the reviewed documents of Tema port expansion as well as interviews with local stakeholders we conclude that – at all hierarchical levels of the *ecosystem based port design hierarchy* – the alternative generation and, hence, the decision making process for the Tema port expansion is primarily informed by economic and functional requirements rather than ecosystem based considerations. Although "harmonization with the environment" is an explicit criterion in the scoring and selection of port lay-out alternatives in the feasibility stage (JICA and GPHA, 2002), the scores on this criterion are hardly underpinned and debatable. The Social and Environmental Impact Assessment (SEIA, SAL Consult Ltd, 2015) does consider environmental impacts, but at that stage most of the planning and design decisions were already made and the number of substantially different design alternatives was limited. As a result, options for avoidance and reduction of environmental impacts may have been missed, and focus shifts to offsetting of environmental impacts.

Although ecosystem based considerations have barely been accounted for in the planning and design documents of Tema port expansion, this does not necessarily imply that ecosystem based alternatives would have been possible at every level of the hierarchy. The Tema case shows that realistic ecosystem based alternatives for the port development (level 1) and site selection (level 2) to meet the Ghanaian transport capacity requirements are barely available in this specific context. Therefore, the design freedom is limited to levels 3 (port lay-out) and 4 (port structures and materials). Our analysis shows that at these levels – from an ecosystem perspective – a wider range of alternatives would have been possible. However, inclusion of ecosystem based alternatives at these levels does not automatically result in more ecosystem based port implementations. Decision making is a multi-stakeholder process requiring trade-offs between different stakes and criteria. Stakeholders may assign more weight to social, economic or functional considerations than ecosystem based considerations. Therefore, ecosystem considerations may eventually not be decisive in the decision making process.

It should be noted that the identification of ecosystem based alternatives at each level of the hierarchy is strongly dependent on the environmental, social and economic context of the port development and, hence, situation specific. Hence, the alternatives identified for the Tema case may not be applicable for port developments elsewhere. Likewise, the characteristics of the Tema case also limit the feasibility of alternative options at the higher levels of the design hierarchy (i.e. level 1 and 2) whereas other case studies may have more potential at these higher levels. As the number of ecosystem based design alternatives is potentially infinite, the examples presented in this paper are by no means complete. The examples provided should be seen as inspirational alternatives to traditional port design rather than a complete overview of design options.

We believe that explicit identification and inclusion of ecosystem based alternatives as part of the decision making process of seaport planning and design, as supported by our framework, is a requirement to arrive at port developments that are (better) harmonized with nature. It may increase opportunities for environmental impact avoidance and reduction, recognition of ecosystem service opportunities otherwise overlooked (e.g. food production, tourism, flood protection) as well as transparent weighing of these opportunities in the decision making process. Although our framework is now not relevant for the implementation of the Tema port expansion anymore, we believe that the lessons learnt can be of added value for planning and design of ecosystem based port developments elsewhere.

ACKNOWLEDGMENTS

This work is funded by the project Integrated and Sustainable Port development in Ghana within an African context (W07.69.206) within the Urbanizing Deltas of the World programme of the Netherlands Organisation for Scientific Research (NWO). This financial support is highly appreciated by the authors. Furthermore, this paper greatly benefited from discussions with and support from Prof. Kwasi Appeaning-Addo and his team from the University of Ghana, Steven Weerts (CSIR, South Africa), Daphne Willems (World Wide Fund for Nature, the Netherlands) and Martijn de Jong and Cor Schipper (Deltares, the Netherlands).

REFERENCES

Alemany, L.J., 2005. Marinas and commercial and fishing ports. Portus 9, 22-27.

Bakermans, B.A., Van der Hout, A.J., Wijdeven, B., 2015. Open ports for container vessels. Proceedings of the 36th IAHR World Congress 2015, The Hague, The Netherlands.

Bruun, P., 1981. Breakwater or mooring system? The Dock and Harbour Authority, Vol XLII, No. 730, pp. 126-129.

Bruun, P., 1990. Port Engineering. 4th Edition, Gulf Publishing Company, Houston.

Bruun, P., 1992. Marine terminal technology. Handbook of coastal and ocean engineering, Vol. 3 – Harbors, navigational channels, estuaries, environmental effects, J.B. Herbich Edt., Gulf Publishing Company, Houston.

Burt, J., Bartholomew, A., Bauman, A., Saif, A., Sale, P.F., 2009. Coral recruitment and early benthic community development on several materials used in the construction of artificial reefs and breakwaters. Journal of Experimental Marine Biology and Ecology 373, pp. 72-78.

De Jong, M.P.C., Weiler, O.M., Schouten, J., 2012. Open water ports – a sustainable designa approach. Third International Engineering Systems Symposium, CESUN 2012, Delft University of Technology, 18-20 June 2012.

De Vriend, H.J., Van Koningsveld, M., Aarninkhof, S.G.J., De Vries, M.B., Baptist, M.J., 2015. Sustainable hydraulic engineering through building with nature. Journal of Hydro-environment Research 9, pp. 159-171.

Dekker, R., Drent, J., Ten Horn, J., McSweeney, N., 2014. Ontwikkeling van mosselbegroeiing op strandhoofden met een Elastocoast® toplaag. NIOZ-rapport, 2014-3. NIOZ: Texel. 33 pp (In Dutch).

Diab, H., Younes, R., Lafon, P., 2017. Survey of research on the optimal design of sea harbours. International Journal of Naval Architecture and Ocean Engineering 9, pp. 460-472.

Dugan, J.E., Airoldi, A., Chapman, M.G., Walker, S., Schlacher, T., 2011. Estuarine andcoastal structures: environmental effects. A focus on shore and nearshorestructures. In: Elliott, M., Dugan, J.E. (Eds.), Treatise on Estuarine and CoastalScience. Elsevier, New York.

ESPO, 2012. ESPO Green Guide – Towards excellence in port environmental management and sustainability. European Sea Ports Organization, Bruxelles, Belgium.

Firth, L.B., White, F.J., Schofield, M., Hanley, M.E., Burrows, M.T., Thompson, R.C.,Skov, M.W., Evans, A.J., Moore, P.J., Hawkins, S.J., 2016. Facing the future: theimportance of substratum features for ecological engineering of artificialhabitats in the rocky intertidal. Mar. Freshwater Res. 67, 131–143.

GPHA, 2017. Tema Port Performance 2000-2016. http://www.ghanaports.gov.gh/publications.aspx (Accessed 7 March 2018).

GPHA, 2018. Tema Port – Our History and Future. Ghana Ports and Harbours Authority. http://www.ghanaports.gov.gh/page/15/Our-History-And-Future (Accessed 7 March 2018).

Gray JS (1997) Marine biodiversity: patterns, threats and conservation needs. Biodiversity & Conservation 6:153-175

Halcrow Engineers PC, 2010. Feasibility Study for the Ghana Ports of Tema and Takoradi. Vol 1-3., New York, April 2010.

Hevner, A.R., Salvatore, T.M., Park, J., Ram, S., 2004. Design science in information systems research. MIS Quarterly, Vol. 28, No 1, pp. 75-105.

JICA and GPHA, 2002. The development study of Ghana sea ports in the Republic of Ghana. Final report. Vol 2-4.

Lam, J.S.L, and Notteboom, T., 2014. The Greening of Ports: A Comparison of Port Management Tools Used by Leading Ports in Asia and Europe. Transport Reviews Vol 34 Issue 2, pp 169-189.

Ligteringen, H., 2012. Ports and Terminals. Delft University of Technology. Delft Academic Press.

Mercader, M. et al., 2017. Small artificial habitats to enhance the nursery function of juvenile fish in a large commercial port of the Mediterranean. Ecological Engineering 105, pp. 78-86.

Mineur, F., Cook, E.J., Minchin, D., Bohn, K., MacLeod, A., Maggs, C.A., 2012. Changing coasts: marine aliens and artificial structures. Oceanogr. Mar. Biol.50, pp. 189–234.

Nebot, N., Rosa-Jimenez, C., Ninot, R.P., Perea-Medina, B., 2017. Challenges for the future of ports. What can be learnt from the Spanish Mediterranean ports? Ocean & Coastal Management 137, pp. 165-174.

Odum, H.T. and Odum, B., 2003. Concepts and methods of ecological engineering. Ecological Engineering 20, pp. 339-361.

Paalvast, P., Van Wesenbeeck, B.K., Van der Velde, G., De Vries, M.B., 2012. Pole and pontoon hulas: An effective way of ecological engineering to increase productivity and biodiversity in the hard-substrate environment of the port of Rotterdam. Ecological Engineering 44, pp. 199-209.

Peris-Mora, E., Diez Orejaz, J.M, Subirats, A., Ibanez, S., Alvarez, P., 2005. Development of a system of indicators for sustainable port Management. Marine Pollution Bulletin 50, pp. 1649-1660.

Perkol-Finkel, F., Hadary, T., Rella, A., Shirazi, R., Sella, I., 2017. Seascape architecture – incorporating ecological considerations in design of coastal and marine infrastructure. Ecological Engineering, In Press Corrected Proof, https://doi.org/10.1016/j.ecoleng.2017.06.051.

Pachakis, D., Libardo, A., Menegazzo, P., 2017. The Venice offshore-onshore terminal concept. Case Studie on Transport Policy 5, pp 367-379.

PIANC, 2011. Working with Nature. PIANC Position Paper. www.pianc.org/workingwithnature.php (Accessed March 7, 2018).

PIANC, 2014a. Masterplans for the development of existing ports. Report N 158

PIANC, 2014b. 'Sustainable Ports' a Guide for Port Authorities. Report N 150.

PIANC, 2018. Site selection and planning for new ports and special marine terminals on greenfield sites – technical guidelines. Report in preparation.

Ray, G.C., 1988. Ecological diversity in coastal zones and oceans. Biodiversity, pp. 36-50.

SAL Consult Ltd, 2015. Environmental and Social Impact Assessment (ESIA) study of port infrastructure development, port of Tema - Revised EIS, Accra, Ghana. June 2015.

Schipper, C.A., Vergouwen, S., De Jong, M., Vreugdenhil, H., De Bel, M., Schasfoort, F., Minderhoud, S. (2015). Port of the future. Exploratory study. Deltares report.

SIGTTO, 1997. Site Selection and Design for LNG Ports and Jetties. Information Paper No. 14, Society of International Gas Tanker and Terminal Operators Ltd, London, United Kingdom.

Slinger, J.H. (Jill); Nava Guerrero, G. d. C. (Graciela) (2016) Engineering: Building with Nature 101x video #08 – Distilling Ecological Design Principles. TU Delft. Dataset. http://dx.doi.org/10.4121/uuid:20576f6c-e439-4a79-abc4-ad13742c7b48 (https://youtu.be/SCK9j2FfwJg)

Tallis, H., Kennedy, C.M., Ruckelshaus, M., Goldstein, J., Kiesecker, J.M., 2015. Mitigation for one & all: An integrated framework for mitigation of development impacts on biodiversity and ecosystem services. Environmental Impact Assessment Review 55, pp. 21-34.

Thoresen, C.A., 2003. Port designer's handbook. Thomas Telford.

Xiao Z. and Lam J.S.L., 2017. A systems framework for the sustainable development of a Port City: A case study of Singapore's policies. Research in Transportation Business & Management 22, pp. 255-262.

Zheng, Z.Z., 2015. Integrated sustainable port design framework development port masterplan. MSc Thesis, Delft University of Technology and National University of Singapore, https://repository.tudelft.nl/islandora/object/uuid%3Ac7aa0ebf-2b14-48db-9cb5-4e71ff196ff3 (Accessed March 7, 2018).

A. LIST OF SITE VISITS, EXPERT INTERVIEWS AND BRAINSTORM SESSIONS

Date	Activity	Experts involved
February 2017	Field trip Tema port	Prof Kwasi Appeaning Addo and team (University of Ghana), Jacob K. Adorkor (GPHA)
12-05-2017	Brainstorm port layout design	Martijn de Jong (Deltares), Tiedo Vellinga (Delft University of Technology, Port of Rotterdam), Heleen Vreugdenhil (Deltares, Delft University of Technology), Cor Schipper (Deltares), Arno Kangeri (Wageningen Marine Research), Poonam Taneja (Delft University of Technology), Cornelis van Dorsser (Delft University of Technology), Wiebe de Boer (Deltares, Delft University of Technology)
15-09-2017	Open interview World Wide Fund for Nature	Daphne Willems (WWF-NL)
27-09-2017	Open interview CSIR	Steven Weerts (CSIR)
24-11-2017	Brainstorm port design hierarchy	Wiebe de Boer, Jill Slinger, Arno Kangeri
08-12-2017	Research Integration Meeting	Wiebe de Boer, Jill Slinger, Arno Kangeri, Tiedo Vellinga, Daan Rijks (Royal Boskalis Westminster N.V.), Mark Koetse (VU University Amsterdam), Liselotte Hagendoorn (VU University Amsterdam), Peter van Beukering (VU University Amsterdam)
12-01-2018	Brainstorm ecological engineering designs	Wiebe de Boer, Jill Slinger, Arno Kangeri

DURME VALLEY RIVER RESTORATION PLAN. MAINTENANCE DREDGING AND REUSING THE SEDIMENT FOR NATURE RESTORATION AND IMPROVEMENT OF SAFETY AGAINST FLOODING

by

Hans Quaeyhaegens, Dorien Verstraete[1], Peter Ratinckx, Toon Goormans, Roeland Adams[2], Cathy Boone[3]

ABSTRACT

The Durme river is a branch of the Scheldt estuary, characterized by fresh water tidal marshes. Along the upstream part several controlled flood areas are located while in the downstream part navigation occurs. The surrounding polders drain towards the Durme during low tide. A lack of a steady upstream discharge, due to the cut-off of the upstream catchment in the town of Lokeren, and absence of regular maintenance dredging, have led to excessive siltation which has compromised these nature, flood control, navigation and drainage functions. The river restoration plan of the Durme aims at revitalizing the river functions as part of the Sigma flood protection programme, devised to protect the Scheldt estuary against storm surges. Several of these areas (Bunt, Klein Broek, Groot Broek) are located along the downstream branch of the Durme. In order to ensure the full functioning and to prevent further sedimentation at these areas, also the upstream part of the Durme river has to be restored.

The Sigma plan defines the construction of flood control areas, reduced tidal areas and areas subject to depoldering, giving space back to the river and restoring wetland and tidal nature. The construction of the dikes around all these areas requires large amounts of sand. Luckily, the sediment of the Durme generally is of sufficient quality to be used as construction material for these dikes.

The Durme Valley River Restoration Plan defines a cross section, aiming at restoring the gravitational drainage of the surrounding polders and revitalizing the silted tidal marshes. The scouring function and volume to maintain the river cross section during low tide is furthermore enhanced by the depoldering and controlled tidal action, and the restoration of the gravitational drainage of the polders, but is also supported by the construction of a pumping station at the cut-off in Lokeren.

European funding was found to reactivate the upstream flood control area 'Potpolder IV', which has become defunct because of the siltation. The project is a pilot in the USAR project (Using Sediment as A Resource, Interreg 2 Seas). The finer nature and higher degree of pollution provides a greater challenge in the upstream part to re-use the Durme sediment as a construction material. A special installation will be designed and used, at the construction site, to separate the fine polluted sediment from the coarser material used for construction of the dikes. Also in this area two pumping stations have been designed to allow the drainage of the catchment of the watercourse 'Lokerenbeek' crossing the controlled flood area.

In the meantime dredging works have been realized in the downstream section of the river, sediment stockpiles have been created to construct the dikes of the Sigma flood control areas situated in the downstream part. The pumping station in Lokeren has been constructed. The works on the upstream part are expected to start in 2018.

The Durme River Valley Restoration Plan is unique in providing and effectively realizing a restoration scheme of an entire river branch of 17 km, both revitalizing nature and flood protection function. The paper is focussing on the sediment management of the Durme River Restoration Plan.

[1] De Vlaamse Waterweg nv, Lange Kievitstraat 111-113, BE-2018 Antwerp

[2] International and Marine Dredging Consultants NV, Van Immerseelstraat 66, BE-2018 Antwerp

[3] Tractebel Engineering NV, Kortrijksesteenweg, 1144 A, BE-9051 Ghent

1. INTRODUCTION

The tidal Durme is a branch of the Scheldt Estuary. It connects the Scheldt to the city of Lokeren. The Durme serves as drainage for the surrounding polders. In the 1960s the upstream catchment was disconnected to divert part of the discharge to the sea canal Ghent-Terneuzen as part of the necessary fresh water supply. This led to increased siltation of the tidal branch. This siltation led to a more difficult drainage of the polders, i.e. reduced gravitational drainage, which in turn increased the need of pumping stations. Other consequences of the siltation were a reduction of conveyance, and hence an increase of the flood risk in the surrounding polders, a reduced navigability, and the loss of ecologically valuable fresh water tidal marshes.

2. THE DURME VALLEY RESTORATION PLAN AS PART OF THE SIGMA PLAN

2.1 The Sigma Plan

The Sigma Plan (**Figure 1**) aims at reducing flood risk by the construction of flood control areas along the Scheldt estuary. The flood control areas are designed in such a way that they are only flooded during storm tide. The flooding occurs through overflow over the embankments, filling the available storage volume. Some of these areas have been conceived as reduced tidal areas. Via an inlet part of the tidal volume enters the area, resulting in an area subject to tide, albeit with reduced range, allowing tidal nature to develop.

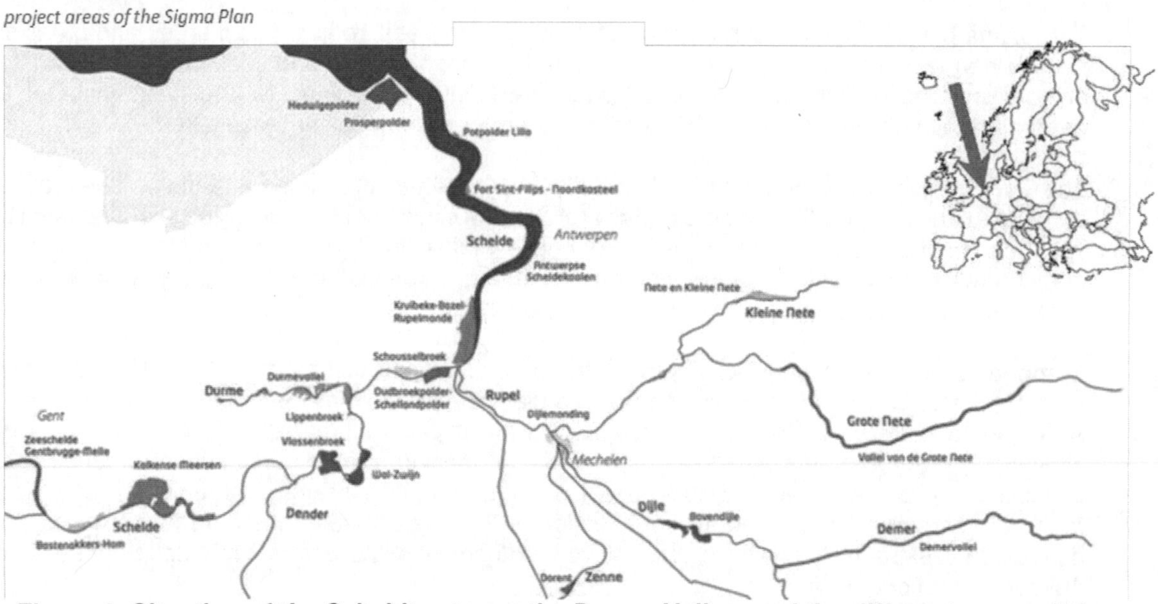

Figure 1: Situation of the Scheldt estuary, the Durme Valley, and the different areas of the Sigma Plan

The flood control areas are designed to protect the Scheldt estuary – hence also the Durme Valley – against flooding, taking into account the future sea level rise. It requires a strengthening and heightening of the current dikes, but also the construction of internal dikes, so-called 'ring' dikes, to protect the hinterland, when the flood control area is being filled by flood water. Also the depolderised areas require the construction of ring dikes to protect the hinterland. It is obvious that the protection of the estuary, and in this particular case also the Durme Valley, requires large amounts of sand, or more in general, granular material.

2.2 The Durme River Valley Restoration Plan

As part of the Sigma plan to protect the Scheldt estuary against flooding, and to restore ecosystem services and nature in the Scheldt estuary, the Durme River Valley Restoration Plan was developed.

This includes the construction of flood control areas, depoldering and the restoration of the river channel and its functions, and rejuvenation of the estuarine nature, partially based on the revalorisation of the sediment (Adams et al. 2013, 2015).

A new cross section for the Durme (**Figure 2**) was designed to accommodate the different requirements of a well-functioning river system:

- The cross section is sufficiently wide and deep to provide the necessary conveyance for drainage of flood water, to partially restore the gravitational drainage, to increase the volume to improve the self-dredging character of the river. Moreover a pumping station was designed and constructed in Lokeren at the former connection with the upstream part of the canal. This pumping station allows flood relief of the upstream catchment, but can also be used to contribute to the sediment management by flushing the tidal Durme.
- The cross section was designed to increase the area of valuable tidal nature. The area of tidal marshes and mudflats lost through works of widening the channel, is compensated by a gain in area by the rejuvenation of the once silted areas in the upstream end.
- In the meantime the widening has to provide the necessary sediment for the purpose of constructing the flood control areas, and to protect the hinterland of the depolderised areas.

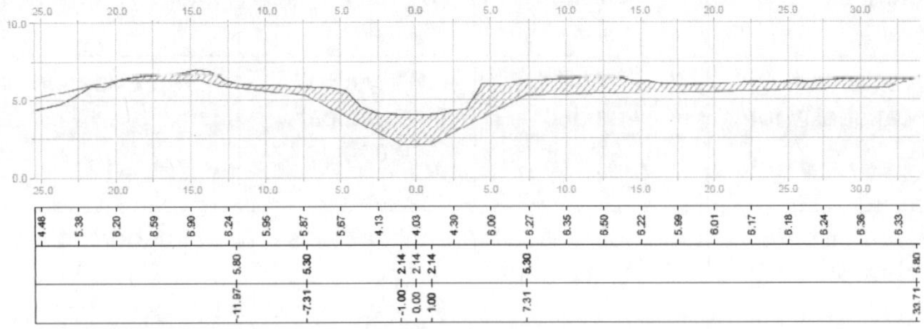

Figure 2: Example of the cross section of the existing and new Durme, showing the increase in section area and creation of suitable surface for rejuvenation of the tidal marshes

3. THE DURME VALLEY RESTORATION PLAN AS A SEDIMENT MANAGEMENT PLAN

3.1 Sediment management within the project

A major challenge is the polluted character of the finer sediment fractions with heavy metals (notably zinc), a consequence of the textile industry once located along the river. As the portion of fines becomes higher upstream, the sediment becomes more polluted. This fairly small grain size is another challenge, requiring separation of the finer material to make the sediment fit for construction.

The need of protection of the estuary was a major driving force behind the realisation of the Durme Valley Restoration Plan. The requirement of construction material (i.c. sediment) for protection of the Scheldt estuary is an excellent opportunity to restore the conveyance, local protection against flooding, and nature functions of the Durme. Moreover the scarcity of sediment for construction purposes in general offers an additional benefit for performing the sanitation of the river.

The Durme River Restoration Plan therefore is more than the restoration of a tidal branch, it is a key project in the Sigma plan for flood protection and nature restoration of the Scheldt estuary. However as a project on its own it basically is a sediment management project. Restoration provides the necessary sediment for the protection against flooding on a larger scale, but offers an opportunity to recreate a healthy river environment on the local scale. The sheer quantity of the involved sediment (2 million m³ of siltation between the nineteen sixties and 2010), the relatively small grain size, the impossibility to directly apply it for construction purposes, and the polluted nature, require specific approaches of sediment management that are discussed in this paper.

3.2 Dredging techniques

Different techniques for different phases in the dredging process have been defined for works in the Scheldt estuary (Santermans & Adams, 2015, Quaeyhaegens et al., 2017), and are also valid in the Durme (**Table 1**). Feasibility is checked with respect to the geographical and tidal constraints, however ultimately it is each contractor's responsibility to propose the final techniques to be used and to prove the feasibility of the proposed procedure.

Table 1: Phases of the dredging process (1-5) and available techniques (A-D)

	A	B	C	D
1. Dredging	Cutter dredger	Hydraulic crane on a pontoon	Long reach	Swamp excavator
2. Transport to the treatment site	Pumping	Hopper barge	Conveyor belt	Transfer to dumpers
3. Treatment site	< 1 km	> 1 km	Slope / Terrain	Directly in dyke body
4. Improvement of chemical quality for re-use	Biological washing	Physical-chemical washing	Immobilisation	Transfer to recycling plant or dumping ground
5. Improvement of geotechnical quality for re-use	Geotubes / Geocontainers	Filter press	Lagooning	Stabilizing

The Durme River Restoration Plan is subdivided in three distinct phases. A Downstream part, a Central Part and an Upstream Part. Each is characterised by specific geographic and tidal conditions, hence the most suitable dredging and sediment preparation techniques can differ for each.

3.3 Treatment of the dredged material

A flowchart (**Figure 3**) was drafted to guide the decision on whether or not the sediment along the Durme, particularly along the upper part, can be used as core material of a new dike, and, if not, what steps are required to make it suitable.

The soil samples taken from the river bed and slope of the dike along the river have been analysed based on the environmental criteria (contamination of the soil sample) and/or the geotechnical criteria (mainly the grain size distribution).

Uncontaminated dredged material determined will be assessed based on geotechnical criteria (grain size criteria). The coarse-grained material taken from the river bed and/or the slope of the existing dike can be directly re-used as core material for the new ring dike.

The fine-grained material (small fractions (I+II) higher than 10%) is assessed based on the plasticity index criteria and/or UK-ADAS texture triangle. Material classified as fine sand or silty to clayey sand can be directly re-used as core material in the new dike.

Other sediments (sandy loam, silty loam ...) require treatment to reach criteria for workability and geotechnical stabilisation. The treatment exists in dewatering of the material or using additives. Once the material has been treated, the sediments can be used as core material.

The exceedance of acceptable Zn, total PCB, Cd, Cr III, Cu, Benzo(a)pyreen and mineral oil concentrations for free use of dredged material (Flemish legislation: VLAREBO Annex V) is a first step of the assessment of treatment requirements. Treatment of the material aims at obtaining the VLAREMA certificate for the use as secondary raw material, based on the allowed total concentrations of heavy metals and a various number of organic compounds and on the leachability criteria for heavy metals. If one of the criteria is exceeded an on- or off-site treatment of the material can be considered to finally make a re-use possible. A mobile treatment plant is considered for obtaining the VLAREMA certificate, as off-site treatment may not be economically feasible. This requires an adapted dredging strategy.

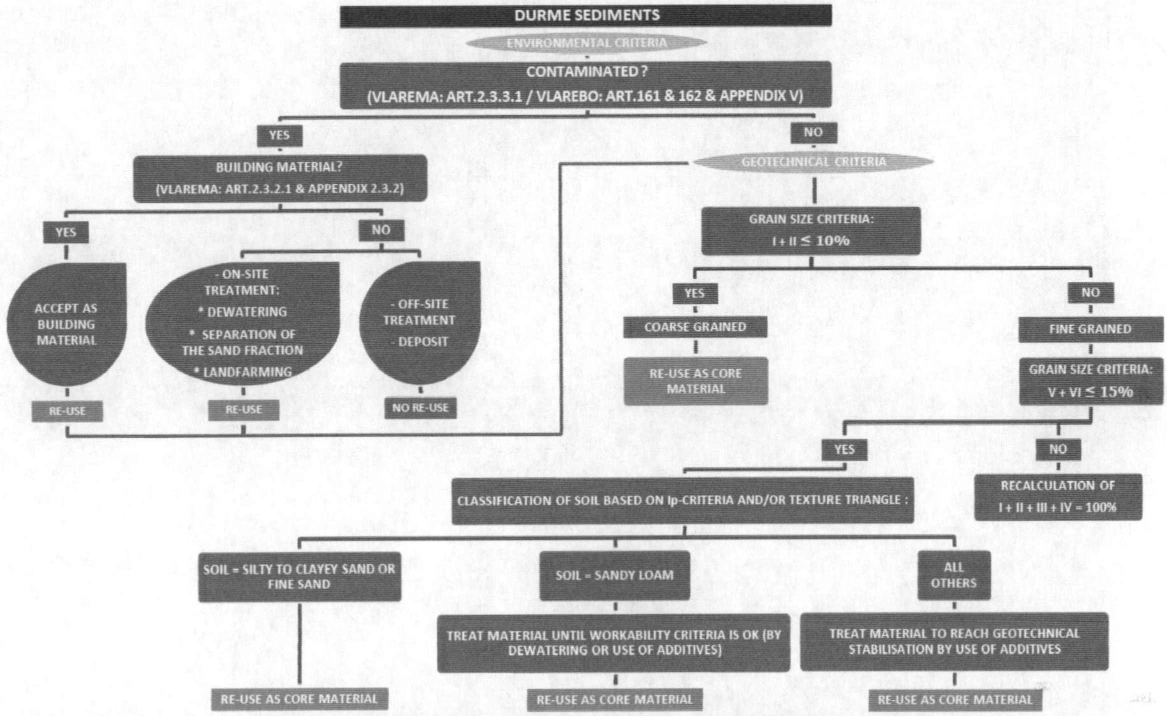

Figure 3: Dredged sediment re-use decision flowchart

4. THE DOWNSTREAM AND CENTRAL PART OF THE DURME

Figure 4: outlet of pipeline in lagoonation area (temporary sand stock) along the Durme

The downstream part of the Durme is the widest, and still navigable today. Dredging started in this area. Cutter suction dredgers were used for dredging the design profile, the sludge being led to basins for lagooning. These basins were realized by constructing the dikes for one of the areas for future depolderisation (**Figure 4**). In the lagooning basin, the coarse material on the one hand settles near the outlet of the pipeline transporting the dredged material. On the other hand the fines reach the far extremity of the basin, allowing to separate the finer material from the sand used for construction.

The fines are removed for processing in disposal areas or sanitation. The sand constitutes a temporary stock for future construction of the Sigma dikes, in both the flood control areas and depolderised areas (Bunt, Groot Broek and Klein Broek).

The central part of the Durme is much narrower, but still wide enough to be accessible for a pontoon, especially after the dredging of the downstream part. Dredging occurred with a hydraulic crane on a pontoon (**Figure 5**), leaving enough space for a barge to be filled for transport to a temporary sand stock for the construction of the Vlassenbroek flood control area, some 10 km upstream from the Durme mouth along the Scheldt river.

The project of the downstream and central part of the Durme Valley River Restoration Plan included the design and execution of the dredging works, and the lay-out of temporary sand stocks for future dike construction.

Figure 5: Hydraulic crane on a pontoon, with a barge, during low tide – Central part of the Durme

5. THE UPPER PART OF THE DURME

5.1 Restoration of the flood control area 'Potpolder IV'

The sediment recovered in the upstream part will be used for restoring the flood control area 'Potpolder IV' (**Figure 6**). Potpolder IV has been serving as flood plain for the Durme river since the nineteen forties. However, the dikes around the flood plain no longer comply with the actual safety levels for flood protection as defined in the Sigma plan. Dredged material from the Durme river will be used as building material for the construction of new embankments. The project is a pilot within the Interreg 2 Seas programme USAR "Using Sediments as A Resource". The project covers the complete design of a new ring dike and the adaptation of the existing dikes along the Durme river to an overflow dike in order to realize a flood control area. Dredged material has been examined and treatment techniques have been proposed to be able to re-use all the material within the new dikes, including contaminated material, in accordance with all environmental regulations. Treatment techniques include dewatering and separation of the fine fraction, stabilization techniques in order to obtain better geotechnical characteristics or the use of geobags or geocontainers in the core of the dikes. The design as well includes all required works to the watercourses in the area, as well as the construction of two new pumping stations. Works are planned to start in 2018.

The flood control area 'Potpolder IV' was initially designed in 1938 and dikes around the area were built in the 1940s and 1950s. Actual crest levels vary between +5,1 and +5,7 m TAW (TAW = the Belgian reference level).

The main purpose of the project is the restoration of the flood control area according to the actual standards by building a new ring dike with a crest level of +8,0 m TAW to prevent flooding beyond the flood area limits, and adapting the existing dike along the Durme to an overflow dike (weir) with a crest level below +6,8 m TAW, in order to obtain efficient filling of the area and an overflow frequency of at least once a year (determined by numerical hydrodynamic modelling). Overall perspective is a better protection of the valley against uncontrolled flooding and lowering the maximum water levels in the upstream part of the Durme river.

Today the actual crest level of the dike along the Durme fluctuates due to the deposition of sediment, causing uncontrolled overflow. The project aims at moving to a controlled regime with an anticipated overflow frequency and controlled flooding and emptying durations. By lowering the crest level and increasing the length of the overflow dike the top of the tidal wave passing through the Durme river can be "captured" and a maximum water level imposed to the river. This way an artificial boundary condition is set for the upstream river stretch and water levels are controlled in a better way. Rainfall runoff conveyed by the local watercourse 'Lokerenbeek' is stored in the flood control area during incoming tides, whereas it is emptied during low tide periods.

Figure 6: The targeted area for restoration of the flood control area 'Potpolder IV' is enclosed between the motorway E17 and the Durme

The project includes the construction of two pumping stations, to allow the drainage of the area upstream the flood control area, and the emptying of the area after flooding of the flood control area.

This discussion will however focus on the investigation whether and how the dredged material resulting from the maintenance works in the upper part of the Durme river can be used within the body of the new ring dike.

5.2 Project characteristics

Dredging in the upstream part of the valley to provide material for use in the new ring dike concerns a volume of around 260 000 m³. The river stretch to be dredged is about 8 km long and is situated between the weir in Lokeren (upstream) and Waasmunster bridge (downstream).

Due to the restricted dimensions of the river and the tidal process limiting the water depth in periods of low tide, the use of a cutter dredger or a pontoon with hydraulic crane is not possible. The accessibility and stability of the existing dikes along the Durme, makes a long reach crane or swamp excavator viable options, the first having the advantage of working outside the actual river bed, the latter being able to be operational in the mud flats but not operational in the river bed at high tide. The choice will also largely depend on the transport facilities from the dredging location to the treatment site.

Transport techniques are a function of distance between dredging and treatment location and the proximity of residential areas. Even if the flood control area is situated more or less in the middle of the (upper) Durme valley, transport distances are large and transport along the winding river bed and dike will be difficult. It is therefore uncertain that pumping techniques may be a feasible solution. The use of hopper barges is excluded due to insufficient river bed dimensions and water depth. Conveyor belts are only feasible in combination with a mechanical dredging (or dewatering) technique but due to the same constraints as mentioned before not an option in this case, leaving only the option to directly transfer to dumpers.

The temporary treatment site will most likely be organized in the flood control area. As treatment will be part of the contractor's responsibility it cannot be excluded that smaller intermediate treatments and/or dewatering sites are set up as well.

Whereas dredged sediments are usually dehydrated by means of lagooning or natural dewatering, techniques that need time and space, De Vlaamse Waterweg aims at finding a technique of "immediate" re-use without intermediate storage for dewatering.

5.3 Sediment evaluation

One of the most important dredging and structural parameters of the sediment with respect to re-use and evaluation of possible treatment procedures is the fines fraction or the content of particles smaller than 63 µm. Between Lokeren (upstream) and Waasmunster (downstream) the fraction of fines varies from 80% to 15% with an average of 45%, with an increase of fines in the river bed from downstream to upstream with a d_{50} of around 50 µm (silt) to 90 µm (fine sand) (**Figure 7**).

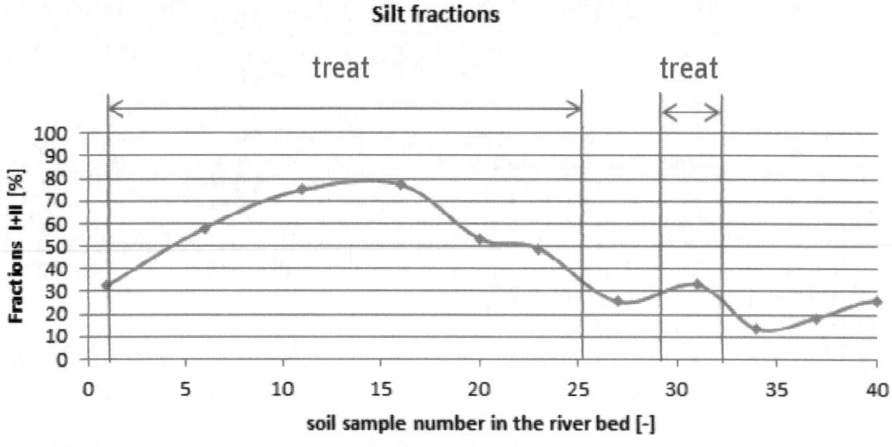

Figure 7: Determination of sediment for treatment, samples 1 to 40 from upstream to downstream

Classification of typical soil samples is shown in **Figure 8**. These classifications will lead to the determination of treatment techniques to be applied in order to make the material appropriate for re-use. Dredged material from locations as classified in the left figure are suitable for immediate re-use in the core of a new dike, whereas dredged material from locations as classified in the right figure require treatment before re-use.

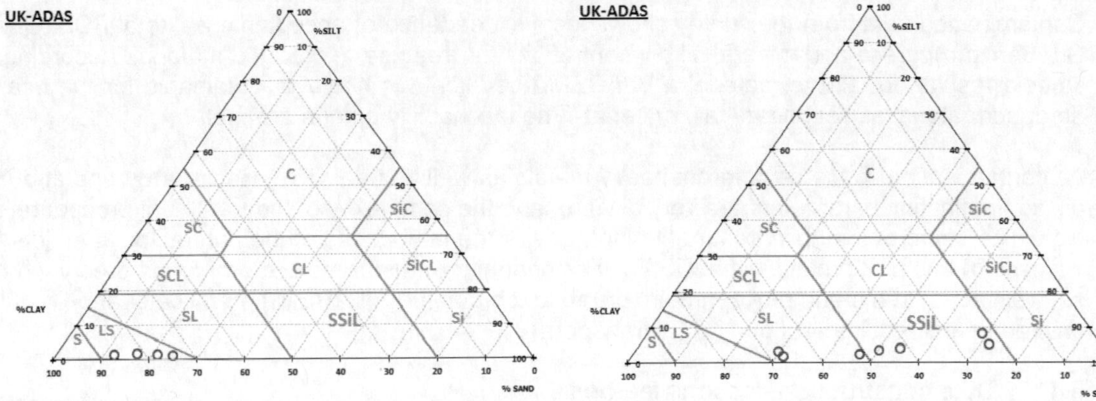

Figure 8: Classification of soil samples in the Durme river bed based on the UK-ADAS textural triangle

The samples were additionally tested for environmental quality characteristics to determine whether the sediment can be used as building material or as secondary raw material. Flemish legislation (VLAREMA, Flemish regulation on sustainable management of material cycles and waste products) defines "building material" as any material which is used within infrastructural works. Material that is contaminated in any way is defined as "waste material". Some specific waste material may switch from the classification 'waste' to 'building material' after treatment to be used for certain applications as recycling or re-use, and if they meet specific criteria:

- The material is used for a specific purpose;
- There is a (market) demand for the material;
- The material meets the technical requirements for the purpose and the concerning legislation and standards;
- The use of the material has no negative impact on the environment whatsoever.

Figure 9 compares the quality standards, green being acceptable as building material, red meaning there is exceedance of certain parameters making the sediment unsuitable for direct re-use. An additional distinction was made between sediment from the river bed and sediment from the bank slopes. It is clear from the figure below that a large volume of dredged material is unsuitable as building material without treatment for re-use in the new dike cores.

Figure 9: Map indicating exceedance of quality criteria for use of the sediment as building material, both in the river bottom (red line) and the slopes (orange area)

Contamination data from the survey shows local exceedance of acceptable Zn, total PCB, Cd, Cr III, Cu, Benzo(a)pyrene and mineral oil concentrations for free use of dredged material. According to the sample results, for all the material a VLAREMA certificate is however obtainable for the use of the dredged material as secondary raw material if the leachability criteria are met.

As contamination concerns different heavy metals as well as total PCB, Benzo(a)pyrene and mineral oil, contamination parameters are very diverse, and the complexity of the treatment procedure(s) very high. The setup of a mobile treatment plant, one of the main tasks of the contractor, is critical for the success of this pilot project within USAR. The contractor must therefore propose a dredging strategy as well as a treatment procedure in order to obtain the VLAREMA certificate and to fulfil the environmental requirement of minimal transport.

5.4 Dike construction: lessons learned

A few kilometres upstream from the confluence of the Durme and the Scheldt river a number of comparable flood control areas have been developed over the last years. One of these projects concerns the Vlassenbroek polder. Part of this polder is transformed into a tidal area with a controlled, reduced tide, whereas the other part will only flood when extreme storm surges occur (Van Nederkassel et al., 2014; Van Zele et al., 2014; Van Nederkassel et al., 2015, Quaeyhaegens, 2017, Quaeyhaegens et al., 2017). First investigations have been made with regard to the re-use of dredged material within the dike core. A number of treatment scenarios have been examined in order to determine the dosage of additives to obtain the required geotechnical properties (stability, permeability) as the same problems with too large amounts of fines were detected here.

However the dredging techniques in the Vlassenbroek project were completely different as the ones required or feasible in the upstream part of the Durme, as the dredging works were carried out in the river Scheldt and the central part of the Durme, making use of barges. The contractor will need to propose an adopted procedure with smaller scale dredging techniques combined with an adequate treatment technique (**Figure 10**) in order to be able to incorporate a maximum amount of dredged material into the ring dike core.

Figure 10: Example of a mobile treatment plant (Vlassenbroek)

The net volume of core material to be incorporated in the ring dike is 160 000 m³, while a dredging volume of 260 000m³ is foreseen. This leaves the possibility to reduce the volume of the dredged material by dewatering and treatment before re-use. In order to give the contractor all the necessary elements to be able to determine the best combination of dredging and treatment techniques, a sounding and sampling survey was planned before the publication of the tender.

6. CONCLUSIONS

In the framework of the Sigmaplan, which aims at reducing flood risk and increasing the ecological value along the Scheldt estuary, the Flemish waterway manager De Vlaamse Waterweg is executing the Durme Valley River Restoration Plan. The restoration plan wants to revitalize the main functions of the river, i.e. drainage of the surrounding areas, evacuation of flood water, navigation, and nature. While the downstream and central part of this plan are already executed, execution of the upstream part is scheduled for 2018.

In the project the focal point is the re-use of dredged material from the tidal Durme for the renovation of the historical flood control area 'Potpolder IV' situated along the river. This renovation involves the construction of a new ring dike with a length of about 3600 m, and the conversion of the existing dike along the river into an overflow dike over a length of approx. 2300 m. The project aims at a 100% re-use of the dredged material. The conditions to dredge are challenging: a small draft, a limited tidal window, and long transport distances. Moreover the nature of the dredged material does not always allow for a direct re-use, neither from a geotechnical nor an environmental point of view. Hence the material will have to be treated.

The paper discussed some possible dredging and treatment techniques. Re-use will be evaluated based on environmental and geotechnical criteria. Dredging strategy and mobile treatment plant setup will be organized in order to obtain the VLAREMA certificate and to fulfil environmental requirements of minimal transport. However, ultimately the choice of techniques (and responsibility) will lie with the contractor (tender to be published at the time of writing).

After successful execution, the Durme river will once again be a well-functioning river, where nature, navigation, and flood protection all have its place.

7. REFERENCES

Adams R., Verhaegen K., De Smedt P., Quaeyhaegens H. (2013) The Durme Valley River Restoration Plan, European River Restoration Conference, September 2013, Vienna

Adams R. et al. (2015). Rivierherstelplan Durmevallei (The Durme valley river restoration plan). IMDC RA13030. Waterwegen & Zeekanaal NV.

Quaeyhaegens H. (2017) How beneficial sediment reuse contributes to improve flood defense in Flanders, Belgium, CEDA Days, Rotterdam November 2017

Quaeyhaegens H., Ratinckx P., Van Rompaey M. and Boone C. (2017) Renovation of a Controlled Flood Area in the Scheldt Estuary using dredged material from the Durme river. CEDA Dredging Days 2017

Ratinckx P., Van Rompaey M., Boone C. and Quaeyhaegens H. (2017) Renovation of a flood area in the Scheldt estuary using dredged material from the Durme river SedNet Genova 2017

Santermans J., Adams R. (2015). Baggertechnieken fase 3 (Dredging techniques phase 3). IMDC NO15326. Waterwegen & Zeekanaal NV.

Van Nederkassel J., Van Zele S., Van Renterghem B., Vermeersch T., Quaeyhaegens H. (2014). The use of engineered sediments for the construction of a compartment dike in the controlled flooding area of Vlassenbroek (Belgium), Proceedings of the South Baltic Conference Dredged Materials in Dike Construction, Rostock, Germany, 87-95.

Van Nederkassel J., Van Zele S., Van Renterghem B., Vermeersch T., Quaeyhaegens H. (2015). The use of engineered sediments for dyke construction in the flood control area of Vlassenbroek. Terra et Aqua, Number 138, 5-16.

Van Zele S., Pallemans I., van Nederkassel J., Vermeersch T., Quaeyhaegens H. (2014). L'emploi de sédiments traités pour la construction d'une digue de compartimentation (Belgique), XIIIièmes Journées Nationales Génie Côtier – Génie Civil, Dunkerque, France, 1063-1076.

New Guidance on Applying Working with Nature to Navigation Infrastructure Projects

by

***Victor S Magar*[1], *Kirsten Wolfstein*[2], *Juan Savioli*[3], *Ellen Johnck*[4], *Sim Turf*[5], *Johny Van Acker*[6], *Paul Scherrer*[7], *Daan Rijks*[8], *Lauren Dunkin*[9], *Danielle Amber*[10], *Pradeep Muguntham*[11]**

ABSTRACT

PIANC's Work Group (WG) 176 on Guidance on Applying Working with Nature to Navigation Infrastructure Projects, was tasked by PIANC to prepare guidance that raises awareness of natural ecosystems, inspires the navigation infrastructure community to embrace natural systems design, and promotes expanded acceptance of the Working with Nature (WwN) approach by providing a selection of case studies to illustrate how WwN applies to navigation infrastructure projects, identifying associated tools, steps, and practices.

INTRODUCTION

WwN is an integrated approach that pursues win-win solutions for navigation infrastructure projects and the environment through careful consideration of natural processes, ecosystem impacts, stakeholder engagement and strategies to maximize opportunities for navigation and nature. Before being implemented, most maritime infrastructure projects must undergo extensive environmental and cultural impact evaluations demonstrating avoidance, minimization, or mitigation of impacts. WwN goes beyond impact avoidance or mitigation and looks for opportunities to integrate nature into project designs, thus creating added value for the project and the environment. As with most contemporary decision-making processes, the WwN philosophy combines social, environmental, and economic considerations into decision making, providing an integrated approach to project development and management. WwN offers an opportunity to design projects with greater holistic integration within the natural environment to secure timely authorizations from regulators.

There are numerous ways in which WwN can be implemented, from altering the surface or texture of submerged structures and thus creating or expanding aquatic habitats, to using natural systems such as islands, marshlands and mangroves, to protecting nearshore environments from severe storm events. The thoughtful implementation of WwN should be aligned with nature, while helping project proponents and environmental stakeholders achieve long-term infrastructure and environmental goals (PIANC 2008, 2011). This Guide for Applying Working with Nature to Navigation Infrastructure Projects discusses WwN in the context of navigation infrastructure, specifically, integrating natural systems and engineering.

For many of those involved with the development and design of waterborne transport infrastructure (e.g., environmental scientists, engineers, project owners and decision makers), WwN is a relatively new paradigm. WwN aims to maximize opportunities to work with natural processes to deliver environmental outcomes that go beyond merely avoiding or compensating for environmental impacts.

[1] Ramboll, Chicago IL, USA
[2] Hamburg Port Authority, Hamburg, Germany
[3] DHI Water & Environment, Malaysia
[4] Ellen Johnck Consulting, USA
[5] Flemish Department of Mobility and Public Works, Belgium
[6] De Vlaamse Waterweg NV, Afdeling Bovenschelde, Belgium
[7] Ministry of Ecology, Energy, Development, France
[8] Boskalis, The Netherlands
[9] USACE-ERDC, Vicksburg MS, USA
[10] Ramboll, Ann Arbor MI, USA
[11] FourPeaks, Chicago IL, USA

While WwN can be contemplated during all phases of a project, it should ideally be considered early in the development of navigation infrastructure projects, when flexibility is maximal. By maintaining a determined and proactive approach from a project's conception through to its completion, opportunities can be maximized and importantly, frustrations, delays, and unnecessary costs can be reduced.

WwN is fundamentally a philosophy determining the way a navigation infrastructure project is developed. It also involves design and project planning, project implementation, ecosystem and performance monitoring, and adaptive management. The adaptive management component recognizes that navigation projects rely upon learning and adaptation to optimize project outcomes, including reducing energy use and protecting the environment.

WwN requires a subtle but important evolution in the project development approach, by moving beyond the conventional approach of minimizing ecological harm to focus on two new goals; aligning project objectives with existing ecosystem functions and identifying win-win solutions that optimize ecosystem, social, and economic benefits.

As economic growth translates into a global increase in trade and the associated need for new and more efficient waterborne transportation infrastructure persists, the knowledge and experience gained from past successes and failures provide a fresh perspective on infrastructure developments. Adhering to national and international environmental legislation requires an understanding of the impact of infrastructure on ecosystems; non-governmental organizations (NGOs), stakeholders, financial institutes and local communities also demand such awareness. Despite significant progress made in recent years, current approaches to designing and constructing infrastructure typically result in efforts to mitigate for environmental loss. Such approaches are not sustainable.

The issue of sustainability is even more pertinent when we consider that world population and economic growth requires increased infrastructure expansion and upscaling, which places continuous and rising pressures on the natural environment. Environmental challenges ranging from local habitat loss, to the transformation of regional ecosystems, to global climate change, require creative solutions that develop infrastructure in the context of natural ecosystems, so that both may be managed effectively.

WwN thinking can inform the delivery of better environmental protections and enhancements while promoting economic development, reducing delays, and leveraging opportunities to provide local communities with highly sought amenity areas, recreational resources and improved landscape habitats.

WwN APPROACH

WwN offers a framework to align new construction, or to rehabilitate existing infrastructure, with natural processes. This approach serves to:

a) enhance ecosystem viability and resilience

b) minimise negative anthropogenic impacts to the environment.

A holistic understanding of ecosystem structures and processes makes it possible to minimise ecosystem degradation and enhance ecosystem functions on a local, regional or watershed scale. Table 1 identifies the environmental, social, and economic benefits that different interest groups could derive from the WwN approach.

Relevance to Navigation Infrastructure

For navigation infrastructure projects, WwN addresses concerns such as river and coastal channelization, dredging and dredged sediment management and management of sedimentary environments, including wetlands, coastal dunes or beaches and barrier islands. WwN can help to limit or offset habitat loss associated with infrastructure development or to design projects to align navigation infrastructure with natural processes in sedimentary environments. In some cases, such alignments can help minimise continuous maintenance dredging requirements, particularly in hydrodynamic environments subject to severe weather events.

Fundamentally, all projects must work with nature to some extent. WwN strategies can reduce energy associated with construction or maintenance (thus providing an opportunity to reduce greenhouse gas emissions), increase habitat functionality and otherwise enhance the short- and long-term delivery of ecosystem services associated with project requirements. Often, there is no single WwN solution for navigation and waterborne infrastructure projects, but rather a range of options with different degrees of environmental enhancement.

Table 1. Project, environmental, and community benefits associated with WwN

Opportunity	Project Proponent	Environmental Interests (Regulators & NGOs)	Community Interests
Environmental	Better environmental integration by avoiding or reducing environmental impacts during design, and reducing mitigation obligations to residual impacts. Improved environment that provides additional ecological services. Improved economics through better understanding and management of the long term environmental effects.	Environmental improvements and increased services. Better understanding of project impacts on the environment and how to leverage the project to improve the environment.	Improvement of the natural environment. Potential for creation of new recreational features linked to the environment.
Social	Better community acceptance of projects.	Possible identification of Win-Win opportunities for stakeholders. Provide better engagement with regulatory authorities, NGOs, the public, and other stakeholders.	Create and identify win-win opportunities. Greater acceptance of port and navigation projects. Less potential for public obstruction if the community understands the public benefits of the project, and how the project protects and possibly enhances environmental features. Better and more engagement of stakeholders, and, through this, improved social cohesion. Better understanding of community interests, leading to better integration of community goals. More active community engagement.
Economics	Reduced risk of project interruptions to protect the environment. Environmental benefits can support a wide range of funding opportunities. Potentially lowered costs, especially during construction and maintenance. More streamlined permitting / approval.	Potential cost reductions through accelerated permitting and regulatory approvals, avoidance of unintended environmental impacts, greater public acceptance. Potential for funding.	Streamlined construction through community involvement. Aligned community and port / navigation interest, leading to more efficient design. Long-term economic gains through improved Ecosystem Services and development of new business opportunities.

WwN objectives are region-specific, site-specific, and project-specific with respect to hydrologic, ecological, economic, and social conditions. Project objectives should be defined in terms of economic benefits and in terms of the ecosystem services that they provide. Those services can be compared to lost services associated with the project work or implementation of the work. Thus, project objectives not only focus on minimising environmental and ecological harm, including environmental impacts associated with construction or with the completed project, but also on preserving or establishing a more natural environment, enhancing habitat functions where possible and improving aesthetics.

WwN Framework

The basic steps of WwN, listed below, are illustrated in Figure 1, and include:

Step 1. *Establish Project Needs and Objectives.* Define desired goals and objectives, evaluate alternative actions and select strategies that are aligned with nature.

Step 2. *Understand the Environment.* Understand natural forces and opportunities to work with nature.

Step 3. *Engage Stakeholders.* The WwN approach supports the United Nations Development Program (UNDP) sustainability goals; to achieve economic growth, social development and environmental sustainability. Stakeholder engagement is essential to achieving these goals.

Step 4. *Project Design.* Prepare initial project proposal and design to benefit navigation and nature. Identify design opportunities to work with nature and explicitly demonstrate how WwN is incorporated into the project design.

Step 5. *Build and Implement.* Implement the WwN design. Implementation also can benefit from the WwN philosophy, to minimise unintended consequences to the environment during construction.

Step 6. *Monitor, Evaluate and Adapt.* Monitoring is a continuous process executed during multiple project phases. Similarly, WwN is a continuous process that requires monitoring and adaptation through corrective actions for continuous improvement.

To allow for future uncertainties, WwN requires flexible decision making that can be refined as outcomes from current and future projects are better understood. WwN involves developing and implementing a management plan that defines the project goals, periodically reviews progress towards those goals and implements corrective actions (and refines the plan) as needed in response to the outcomes of environmental monitoring.

WwN encourages consideration of site specific ecosystem characteristics during the project design phase, to achieve project objectives for the development, expansion or growth of ports and navigable waterways. Ideally, WwN requires development of a fully integrated approach with input in to the development of project objectives, and before design begins. While it is possible to implement WwN at virtually all phases of a project, incorporating WwN applications during conception, design and early implementation, provides the most promising opportunities to affect positive outcomes for the environment. Greater effort is generally needed to introduce WwN concepts later in the design process, and those efforts may not be as effective as when WwN is introduced during project initiation.

Content of WG176 Guide on WwN

The Guide for Applying Working with Nature to Navigation Infrastructure Projects is organized as follows:

Chapter 1, *Introduction*, introduces the WwN vision, and identifies the intended audience of the report, and provides considerations for economies in transition and developing countries.

Chapter 2, *Background*, provides background information on the WwN approach. The overall framework is introduced, along with beneficiaries of the WwN approach and potential challenges that project owners may face when implementing WwN—by recognizing opportunities and challenges, the hope is that project owners can proactively pursue WwN opportunities for their projects.

Chapter 3, *Context*, describes the context of when and where WwN may be implemented for port and navigation projects. Ports, waterways, and shoreline developments have traditionally relied on conventional infrastructure methods using dredging, steel, concrete or stone for armouring and shoreline protection. WwN recognizes the opportunity to proactively integrate nature into large infrastructure projects to protect or improve natural habitat.

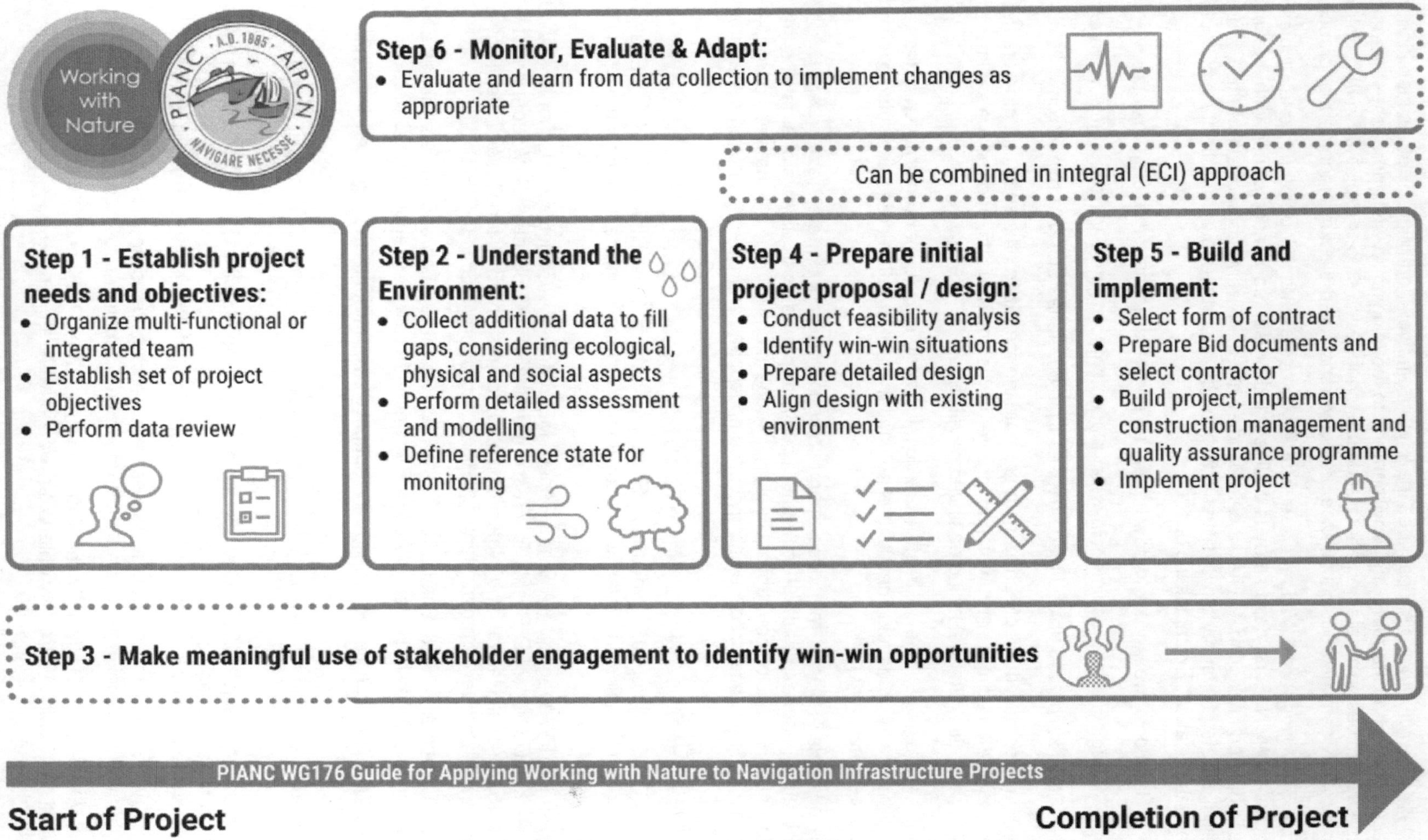

Figure 1. The Working with Nature Framework

Chapter 4, *WwN Framework*, describes the WwN framework, outlining six steps identified in the WwN process, from understanding the environment to developing WwN approaches, and from public engagement to design, implementation, and monitoring. Each step is explained in the context of WwN, and how WwN can and should influence project visions and project management approaches.

Section 5, *Working with Nature Case Studies*, provides 12 independent case studies, providing examples of WwN. Each case study is outlined and presented in accordance with the six-step WwN process defined in Section 4. The 12 case studies are an integral component of this Guide. The case studies illustrate the variety of ways that WwN can be integrated into infrastructure projects to protect and enhance nature. Examples include:

- Creation of wetlands and new habitat as part of large infrastructure projects
- Strategic dredged sediment in-water placement to create a variety of new in-water habitats
- Stabilization and upland placement as fill to create land for port expansion
- In-river placement of sediment to promote the formation of islands that promote natural and targeted sediment accumulation, create new island habitats, and reduce sedimentation in the navigation channel by increasing river velocities
- Leveraging of a new infrastructure development (e.g., tunnel construction across a waterway) to develop new shoreline habitat and recreation areas, including wetlands, beaches, and parks

Each example reflects the cooperation among project owners (e.g., ports), governmental organizations, nature-based NGOs, and the public to promote sustainable alternatives that restore or create beneficial ecological habitat, improve dredge management alternatives (particularly through beneficial sediment use), reduce energy consumption, and improve stakeholder engagement.

REFERENCES

Bridges, T. S., Lillycrop, J., Fredette, T., Suedel, B., Banks, C. J., Russo, E. (2014). Engineering With Nature for Sustainable Water Resources Infrastructure. Technical article for the 33dd PIANC world congress.

Bridges, TS, PW Wagner, KA Burks-Copes, ME Bates, ZA Collier, CJ Fischenich, JZ Gailani, LD Leuck, CD Piercy, JD Rosati, EJ Russo, DJ Shafer, BC Suedel, EA Vuxton, and TV Wamsley. (2015) Use of Natural and Nature-Based Features (NNBF) for Coastal Resilience. Engineer Research and Development Center (ERDC). Vicksburg, MS. ERDC SR-15-1.

Burdon, D., Atkins, J.P. & Bhatia, N., 2011. Ecosystem services and societal benefits provided by estuarine saltmarsh. Report to ABP & EA, December 2011. (Report No. SBB338-F1-2011).

CEDA (2015). Position Paper – Integrating adaptive environmental management into dredging projects - March 2015

De Vriend, H.-J., Van Koningsveld, M. (2012). Building with Nature: Thinking, acting and interacting differently. Eco Shape, Building with Nature, Dordrecht, the Netherlands.

European Commission (1992). Council Directive 92/43/EEC of 21 May 1992 on the conservation of natural habitats and of wild fauna and flora. http://ec.europa.eu/environment/nature/legislation/habitatsdirective/index_en.htm

MA (2005). Millennium Ecosystem Assessment – Ecosystems and Human Well-being. Synthèses. http://www.millenniumassessment.org/documents/document.356.aspx.pdf.

PIANC 2011. PIANC Position Paper 'Working with Nature'. Published October 2008; Revised January 2011. Available at http://www.pianc.org/workingwithnature.php.

Disaster Prevention Facilities and Marine Environment Improvement Effect

by

Naozumi Yoshizuka[1], Hiroshi Matsushita[2], Takashi Nakanishi[3], Hirokazu Nishimura[4] and Kouhei Oguma[5]

ABSTRACT

Concrete structures are indispensable for disaster prevention. Nevertheless, its inorganic structure was sometimes pointed out as a symbol of environmental destruction. In Japan, concepts of ecological consideration were incorporated into Environment Basic Act, present River Act and Seacoast Act in the 1990s, and shapes of concrete structures have been modified. However, many of these were mitigations against development activities and consideration for the environment during development. Under these circumstances, the authors have developed SUBPLEO-FRAME (SPF) combining stone material with concrete and Environmentally Active Concrete (EAC) using amino acids, aiming at compatibility between "disaster prevention" and "environmental conservation". In this paper, examples of realizing conservation and improvement of surrounding marine environment while realizing disaster prevention function are introduced.

1. INTRODUCTION

Concrete structures such as breakwaters, revetment and wave-dissipating blocks are indispensable to protect the functions of ports and harbours as well as in terms of disaster prevention. Nevertheless, its inorganic structure was sometimes pointed out as a symbol of environmental destruction. Installing a structure has been often regarded as a confrontation between development and environment since it has a great influence on the animals and plants that exist there conventionally.

In Japan, since in the 1990s, the concepts of ecological consideration were incorporated into Environment Basic Act, present River Act and Seacoast Act, shapes of concrete structures have been improved. In addition, artificial tideland and seaweed bed creation are also being tried. However, most of these activities are not primarily aimed at regeneration or creation of the environment. These were mitigation of environmental influences due to development activities such as effective use of dredged soil and reclamation, and consideration for the environment during development activities. In recent

KEYWORDS : Disaster Prevention, Marine Environment, SUBPLEO-FRAME, Environmentally Active Concrete, 2030 Agenda, Sustainable Development Goals (SDGs)

[1]Project Engineer, International Division, NIKKEN KOGAKU CO. LTD., 17 F Nittochi-Nishishinjuku bldg. 6-10-1 Nishi-shinjuku, Shinjuku-ku, Tokyo 160-0023, Japan. yoshizuka@nikken-kogaku.co.jp

[2]Chief Engineer, International Division, NIKKEN KOGAKU CO. LTD.

[3]Consultant, NIKKEN KOGAKU CO. LTD.

[4]Section Manager, NIKKEN KOGAKU CO. LTD.

[5]Researcher, Engineering Division, NIKKEN KOGAKU CO. LTD.

years, efforts to further advance these concepts and consider and implement the disaster prevention and the environmental consideration with the same weight have begun. In the Basic Program for the Conservation of the Environment of the Seto Inland Sea 2015 and the Technical Standards for Port and Harbour Facilities in Japan 2018, it is clearly specified that when developing disaster prevention facilities, the environment conservation must be provided.

These trends became clear even from a global standpoint, and '17 Sustainable Development Goals (SDGs)' were established in "The 2030 Agenda for Sustainable Development" adopted at the United Nations Sustainable Development Summit in September 2015. These 17 goals include "Goal 13. Take urgent action to combat climate change and its impacts" and "Goal 14. Conserve and sustainably use the oceans, seas and marine resources for sustainable development". The harmonization between these two goals, that is, development of the construction to achieve both "disaster prevention" and "environmental conservation" is an important key for the realization of a sustainable society.

From this kind of circumstance, our research and development group has developed SUBPLEO-FRAME (Hiraishi et al., 2011 ; Matsushita et al., 2013) combining stone materials with concrete and Environmentally Active Concrete (Sato et al., 2011) using amino acids, aiming at compatibility between disaster prevention and environmental conservation. In this paper, examples of realizing conservation and improvement of surrounding marine environment while realizing disaster prevention function are introduced.

2. EXAMPLES OF SUBPLEO-FRAME COMBINING STONE MATERIALS WITH CONCRETE

2.1 SUBPLEO-FRAME

SUBPLEO-FRAME (hereinafter referred to as SPF) was proposed by Hiraishi et al. (2011) as a new reinforcing method for harbour side of caisson breakwaters against accidental tsunamis and waves (see Fig1). Fig.2 show the image of installation of SPF. It is a simple structure in which concrete frame blocks are installed in the harbour side of breakwater and filling stones are filled into the hole (see Fig.3). The sliding resistance of the caisson breakwater increased by interlocking between filling stones restrained by the frame block and mound stones, which makes it possible to build the resilient structure. Thus far, Matsushita et al. (2013) and Matsushita et al. (2014) have verified the effects on tsunamis and waves by conducting hydraulic model experiments (see Fig.4). The size of SPF is 3.0m long x 3.0m wide x 1.5m high, and the size of the inner frame is 1.8m long x 1.8m wide x 1.5m high. The mass of the single filling stone is around 30kg. Normally, when using stones of around 30kg in the sea, though they will be scattered by overtopping-waves and flows, they can stay there by the restraining effect of the block because the filling stones of SPF are inside the frame block. Also, since there are many voids and roughness on the surface by using stone materials, it is possible to generate the turbulent flow on the surface even if there is only a slight flow in the sea. Therefore, it is difficult for suspended solids such as silt floating in the sea to accumulate, and habitats for various benthos can be created.

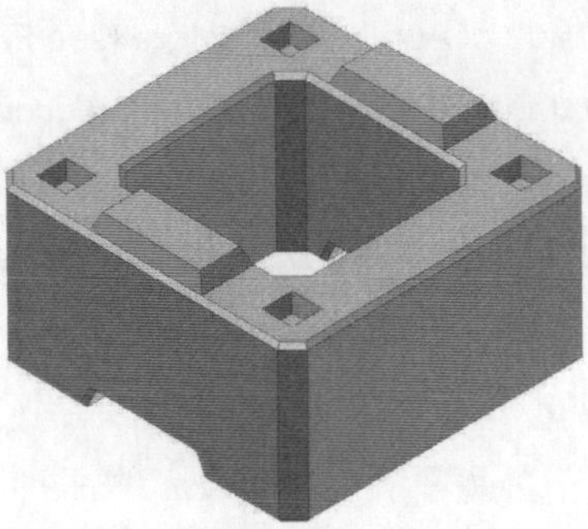

Figure 1 : SUBPLEO-FRAME(SPF)

Figure 2 : Installation Image of SPF

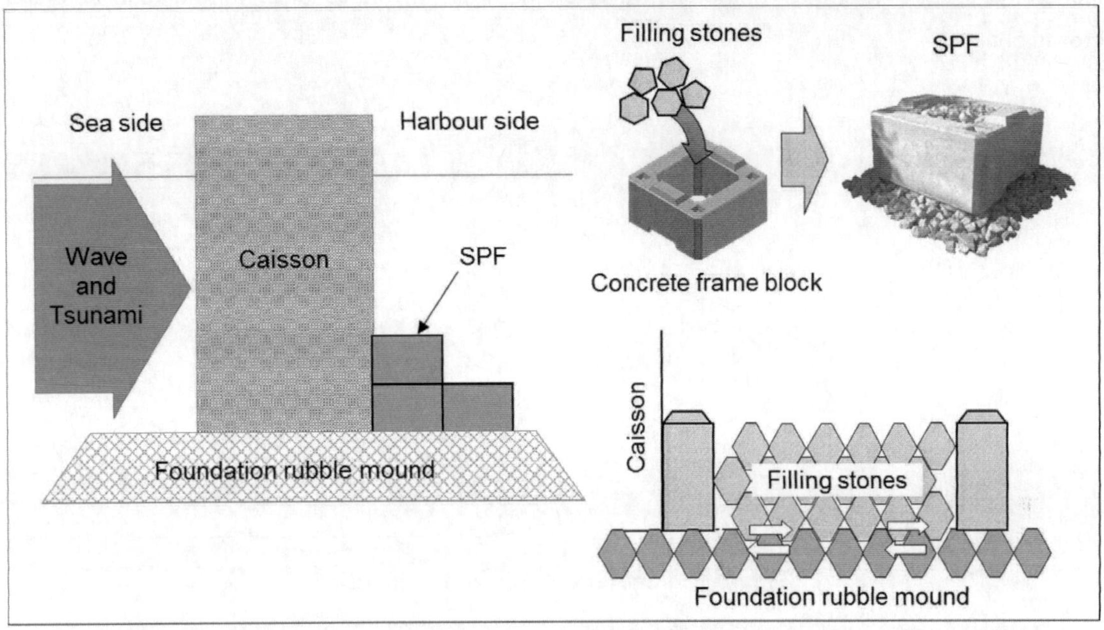

Figure 3 : Structure of SPF

(a) Test against waves in wave basin

(b) Test against Tsunami in wave flume

Figure 4 : Hydlaulic experiment of SPF

2.2 Condition of SPF after the installation in Port of Hachinohe

Port of Hachinohe is a large port located on the Pacific Ocean side of Aomori Prefecture (see Fig.5). The north breakwater was seriously damaged by the Great East Japan Earthquake Tsunami occurred on March 11th, 2011. The breakwater is presumed to have collapsed due to the scouring of the inside the harbour mound caused by the tsunami overtopping discharge along with a lack of sliding resistance of the caisson generated by the force of a huge tsunami. Therefore, SPF was adopted as resilient structure reinforcing the sliding resistant of breakwaters. Fig.6 shows the breakwater immediately after the disaster and after restoration. The SPF is installed in the section of approximately 700 meters inside the harbour of the breakwater. Fig.7 shows an example of a cross-section view. The restoration was completed in June 2013, about five years have passed. In this site, the SPF and foot protection block with flat surface are installed adjacent to each other. The submerged depth of crown of the SPF and the foot protection block is about -7.5 meters, which is a relatively severe condition for the adherence and growth of algae.

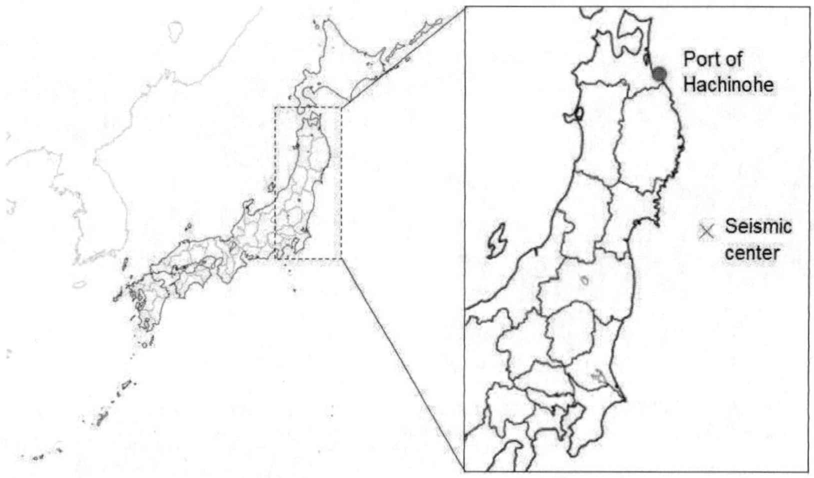

Figure 5 : Location of Port of Hachinohe

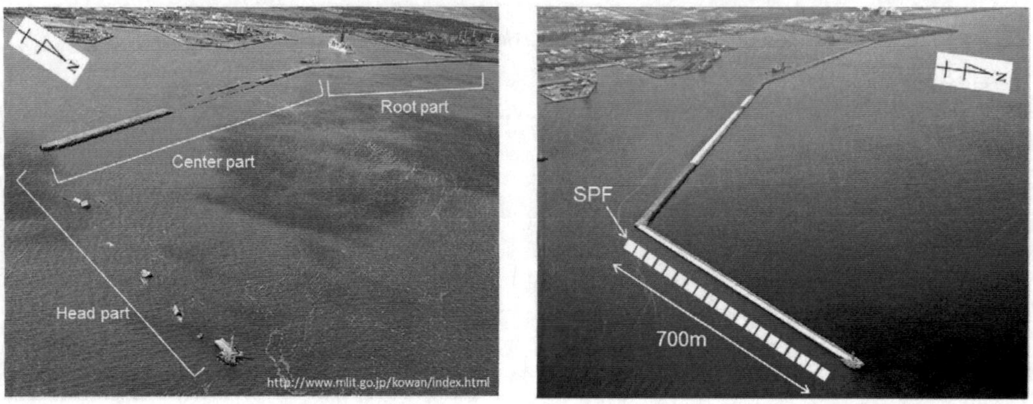

(a) Immediately after the disaster (b) After restoration

Figure 6 : Port of Hachinohe breakwater

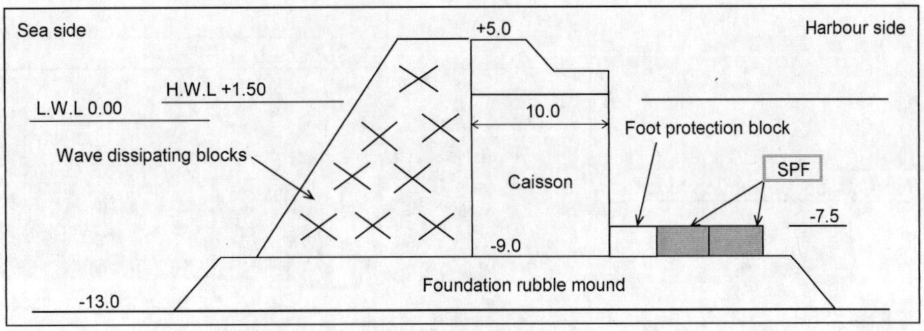

Figure 7 : An example of Cross-section

The monitoring surveys of the condition after the installation of the SPF and the foot protection block were conducted in July 2015, the first survey approximately 2 years after the installation, and in March 2016, the second survey 2 years and 9 months after the installation. Fig.8 shows the conditions of the SPF and the foot protection block. As seen from the photograph of the first survey, suspended solid is accumulated in the foot protection block on the left side, and no animals or plants were observed. For the SPF on the right side, large algae such as Wakame seaweed flourished on the concrete frame, and sea cucumber also inhabited. Large algae also flourished in the filling stones. It is shown that SPF is favorable environment for animals and plants. Subsequently, looking at the photograph of the second survey, large algae could be confirmed slightly on the foot protection block of the left side. On the contrary, inhabiting of large number of large algae and sea animals such as sea pineapple which is a special product of Sanriku District where the port of Hachinohe is located was confirmed on the SPF of the right side, and it was very appreciated by local fishermen. Fig.9 is the pattern diagram of foot protection block and SPF. The characteristics of each are shown as follows;

Foot protection block

I. Suspended solids are easy to accumulate because of the flat surface.
II. Seaweed spores are difficult to survive by the accumulation of suspended solids.
III. Sea animals are hard to inhabit by the accumulation of suspended solids.

SPF

I. Suspended solids are hard to accumulate because of the voids and roughness on the surface.
II. Seaweed spores adhere and grow.
III. Adhering animals (e.g. sea pineapples) inhabit and feed, and filter the suspended solids (water purification effect).
IV. Benthic animals (e.g. sea cucumber) live, and feed and digest sediments (sediment improvement effect).

After the installation of SPF, there was no scattering of the filling stones as confirmed by the hydraulic model experiment, even though it was attacked by several huge typhoons. In addition, SPF has a height of 1.5m, and can be piled with 2 steps, 3 steps and used even in the deep water. This survey showed that SPF is a construction method combining functions of disaster prevention (physical resiliency) and environmental improvement (creation of growth and habitats for living organisms). At the same time, improvement of fishing ground productivity is also expected.

Figure 8 : Results of monitoring survey

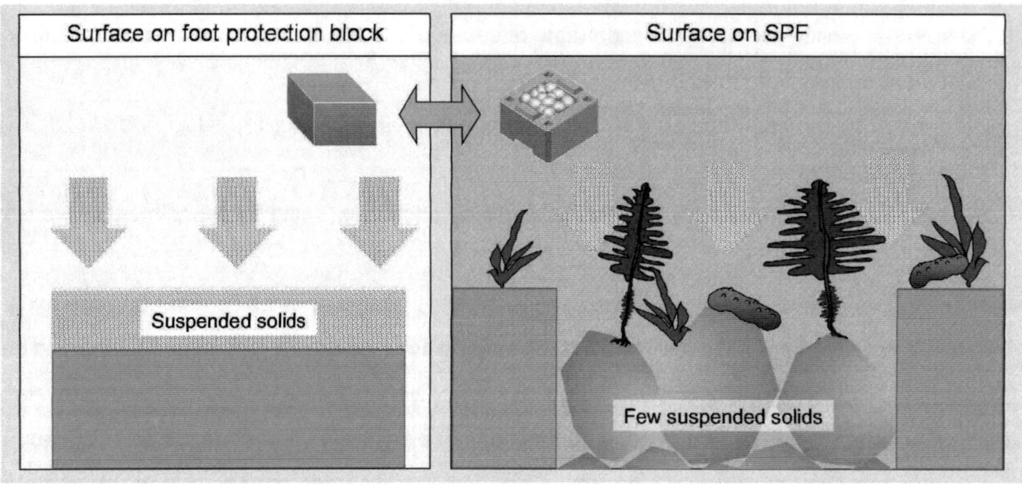

Figure 9 : Diference between foot protection block and SPF

3. EXAMPLES OF ENVIRONMENTALLY ACTIVE CONCRETE USING AMINO ACID

3.1 Environmentally Active Concrete

Environmentally Active Concrete (hereinafter referred to as EAC) is new concrete containing amino acid, developed by Sato et al.(2011) and Nishimura et al.(2014). As a result of conducting concrete strength tests with various kinds of amino acids, Arginine which has less effect on concrete strength was selected. Fig.10 shows the transition of compressive strength of arginine-mixed EAC. The weight of Arginine containing in EAC is recommended 3% of the cement weight ratio. Neither reduction of abrasion resistance by mixing Arginine nor neutralization of concrete were observed.

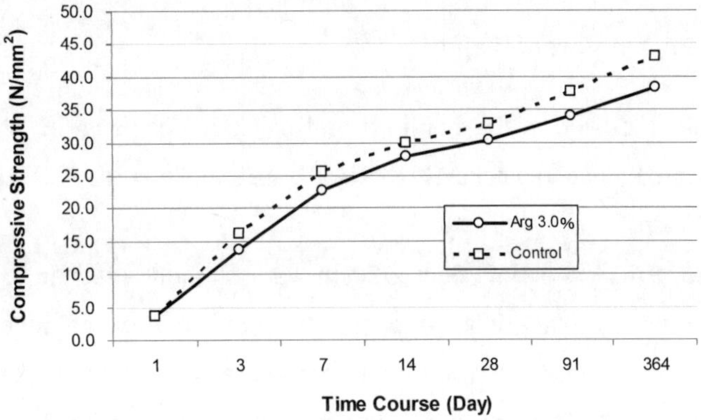

Figure 10 : Temporal Change of Compressive Strength

Fig.11 shows the adhesion state of algae on the specimens made by ordinary concrete (hereinafter referred to as Ord-con) and EAC 8 months after installation in the experiment at actual sea area conducted at the Kojima fishing port in Osaka Prefecture. The amount of arginine added to EAC specimen is 1%, 3% and 5% respectively by cement weight ratio. Many adherent algae flourish on the blocks with high content of Arginine. Fig.12 shows the transition of amounts of chlorophyll-a showing the amount of adherent algae per unit area of the surface of EAC block in which arginine is added 5% by cement weight ratio, and Ord-con block. The amount of chlorophyll-a per unit area of the EAC block was 2-5 times higher than that of the Ord-con block. However, the consumption of algae by sea animals often affects on the data obtained in the experiment at actual sea area. As a result of a laboratory experiments which are not affected by sea animals, the amount of algae on the EAC block is 5 to 10 times more than that of on the Ord-con block (see Fig.13).

Figure 11 : Monitoring condition of blocks taken from underwater after 8 months

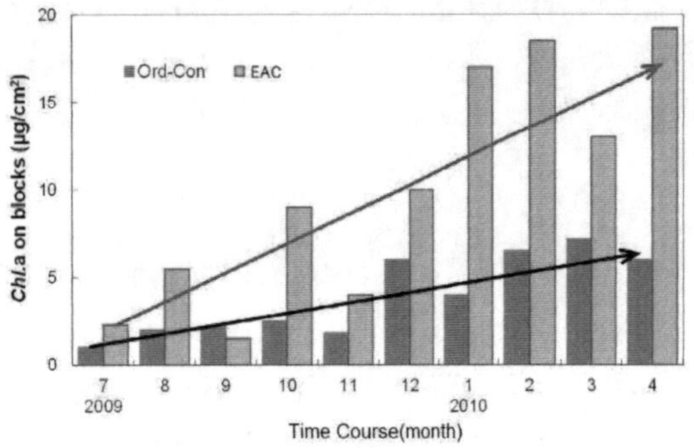

Figure 12 : Time variation of Attached *Chl*.a on blocks in the experiment at actual sea area

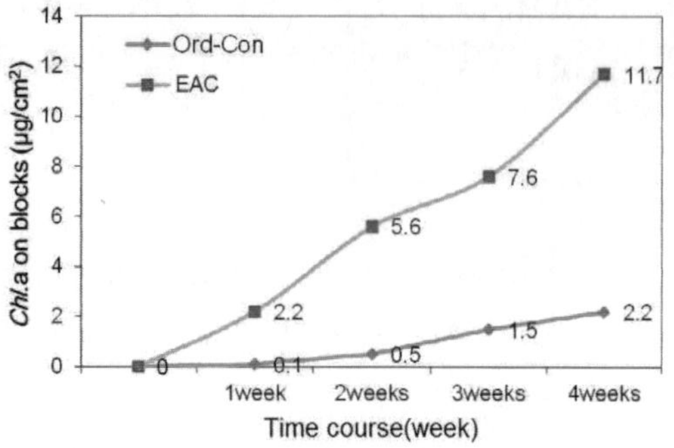

Figure 13 : Time variation of Attached *Chl*.a on blocks in Marine Lab facility

3.2 Example using EAC in Port of Wajima

In the port of Wajima located in the northern part of Noto Peninsula in Ishikawa Prefecture, the breakwater using wave-dissipating blocks, RAKUNA-IV, are installed (see Fig.14). The size of the wave-dissipating block is 20 ton.

As shown in Fig.15, EAC panels of 30cm x 30cm were strongly attached to the legs of the wave-dissipating blocks with adhesive agent and anchor. The amount of Arginine added to the EAC panels is 3% by cement weight ratio. Since the water depth of the breakwater installation location exceeds -10 meters, 5 blocks attached with EAC panels are installed at the position up to -3 meters where the sunlight certainly reaches. A total of 18 EAC panels were attached to each wave-dissipating block.

The blocks attached with EAC panels were installed in July 2013. The monitoring surveys for the effect on ecosystems were conducted 7 times in total.; 1st survey: 1 month after, 2nd survey: 3 months after, 3rd survey: 10 months after, 4th survey: 15 months after, 5th survey: 23 months after, 6th survey: 28 months after, 7th survey : 33 months after. Fig.16 are the photographs of the algae growing on the EAC panels. (a) is the EAC panel in the first survey one month later, it clearly shows that microalgae are flourishing on the EAC panel, compared with surrounding Ord-cons. (b) and (c) are 10 months and 15 months later, and algae are flourishing steadily over time. (d) and (e) are sea urchins and turban shells

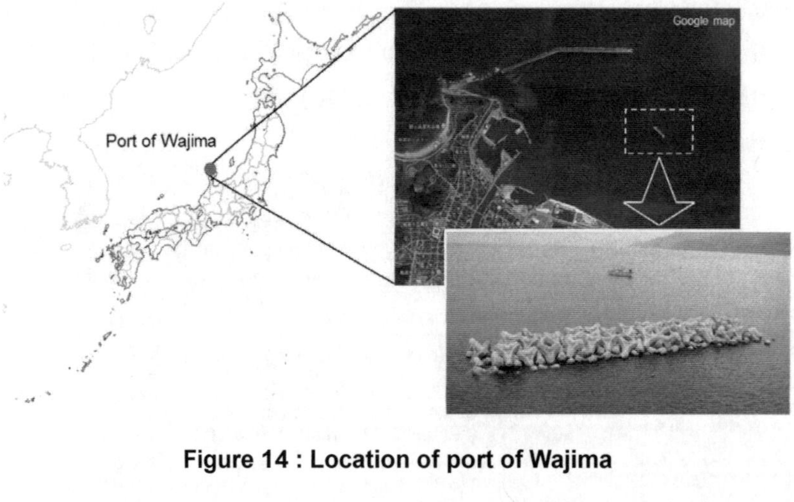

Figure 14 : Location of port of Wajima

gathering on the EAC panel, and (f) is fish migrating around the block. As just described, the stationary fish and shellfish and their fry around the EAC panel, and migratory fish around the block were confirmed. This indicates that the breakwater functions as a habitat for diverse fish and benthos. Fig.17 shows the time variation of the EAC panel. Microalgae flourish for about half a year after installation, and as time passes, it can be seen that the seaweed beds are formed by large algae. Thus by using the EAC panel, since the entire breakwater becomes an artificial fishing grounds and favorable seaweed beds, it is revealed that it is possible to realize both disaster prevention and the environment.

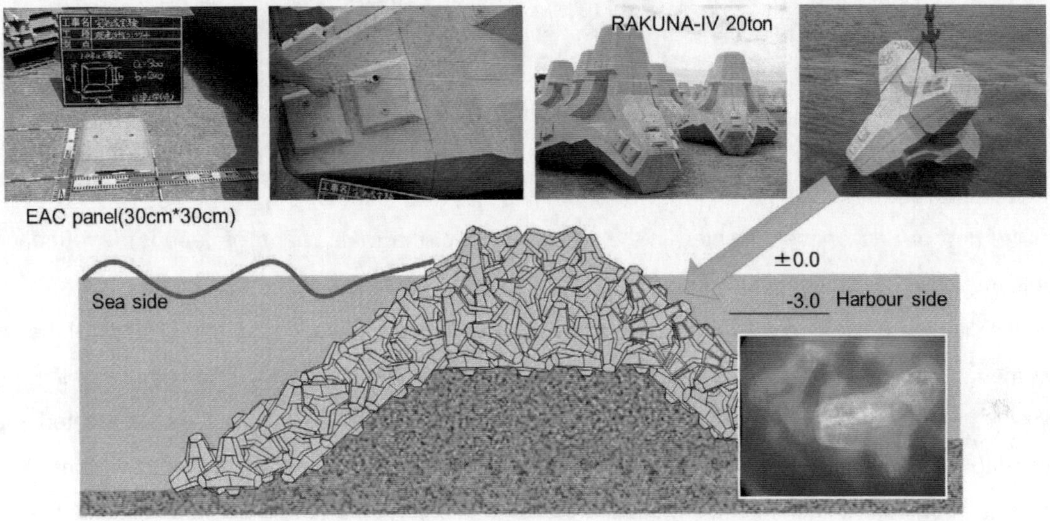

Figure 15 : Wave-dissipating block attached with EAC panel and Cross-section

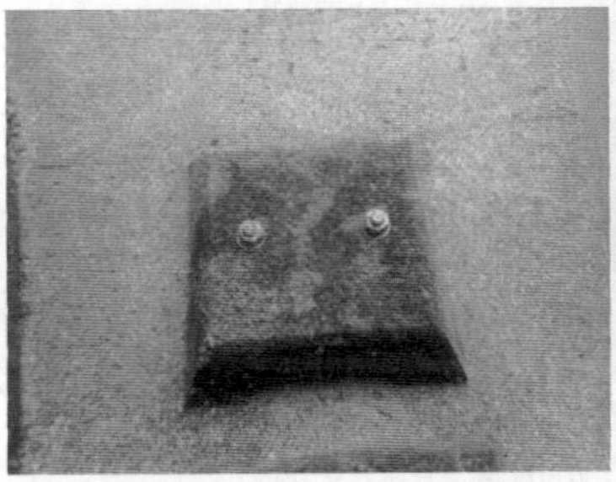

(a) Microalgae on EAC panel
(1 month after installation)

(b) Seaweed on EAC panel
(10 months after installation)

(c) Seaweed on EAC panel
(15 months after installation)

(d) Sea urchin
(23 months after installation)

(e) Turban shell
(33 months after installation)

(f) Fish
(33 months after installation)

Figure 16 : Results of monitoring survey

Figure 17 : Time variation of EAC panel

3.3 Example using EAC in Port of Shimonoseki

In the port of Shimonoseki located in the western end of the main island of Japan, the foundation rubble mound of revetment is armoured by armour blocks, STONE-BLOCK (see Fig.18).

In this site, the EAC panels are attached to the armour blocks installed in the end of slope of the revetment. The Ord-con panels were also used for comparison. Fig.19 are the photographs of the blocks attached with the EAC panels and cross-section view on site. The EAC panels are located at a water depth of -1.0m. The first monitoring survey for effect on ecosystem was conducted in June 2017, 9 months after the installation, and the second survey was conducted in October 2017, 13 months after the installation. The upper row of Fig.20 shows the result of the first survey and the lower row shows the result of the second survey. In the both surveys, it can be seen the growth effect by Arginine

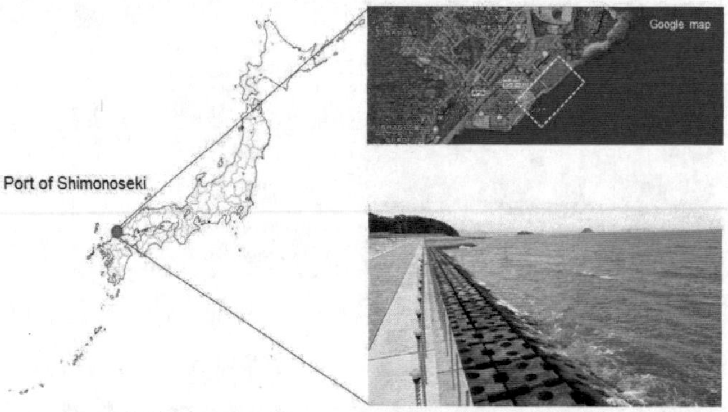

Figure 18 : Location of port of Shimonoseki

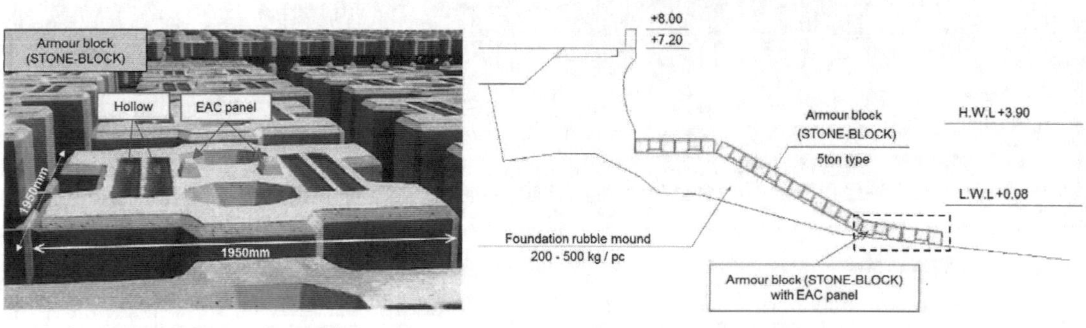

Figure 19 : Armour blocks attached with EAC panels and Cross-section

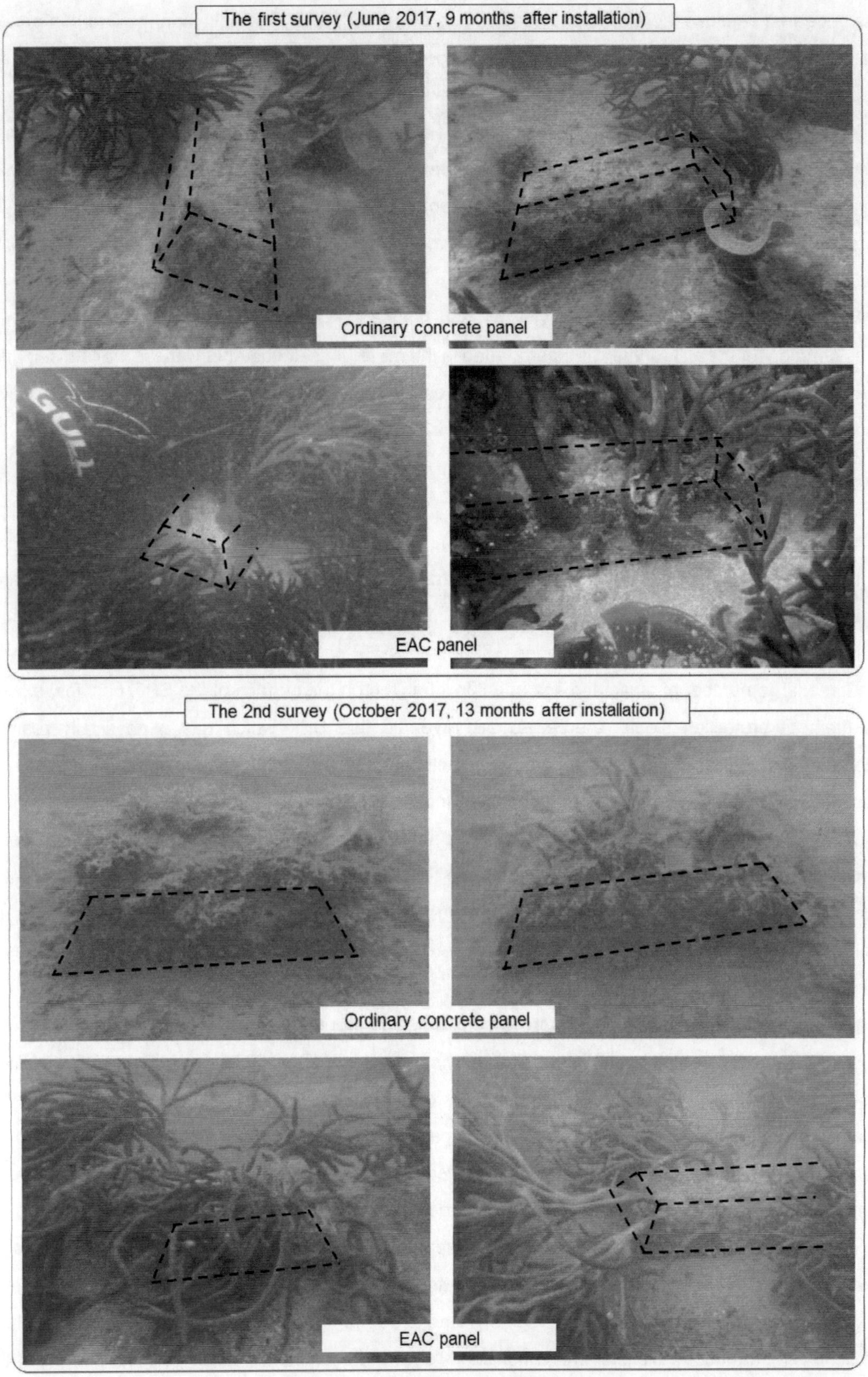

Figure 20 : Monitoring results

has appeared because there are more seaweeds growing on the EAC panels than the Ord-con panels. Mainly growing seaweed is a perennial Codium fragile, which grows most in japan from June to October. Due to its seasonal rise and fall, Codium fragile growing on the Ord-con panels had almost disappeared in the second survey in October. On the contrary, it can be confirmed that many Codium fragile remain on the EAC panels even in such conditions. Therefore, it became clear that the EAC panels are being favorable habitats for living organisms also in the port of Shimonoseki. In the future, as growth of algae spreads throughout the submerged part of revetment, it is expected that favorable habitats will be created for fish as well.

As in the examples of the port of Wajima and the port of Shimonoseki, by using EAC in which consideration for the environment is incorporated in the material itself, together with the wave-dissipating blocks and armour blocks, it is possible to achieve both disaster prevention and the environment as well as further improvement of fishing ground productivity.

4. Summary

In this paper, the examples which realized environmental improvement in the surrounding sea area while demonstrating the disaster prevention function are introduced.

In the example of the port of Hachinohe where using SPF combining stone materials with concrete, though a large number of algae and sea animals inhabited on the surface of the SPF, only few existed on the foot protection block. Therefore, it is revealed that SPF which has a physically resilient characteristic against storm surge accompanying climate change such as sea level rise and abnormal weather and tsunami caused by earthquake, and creates growth and habitat of organisms is a construction method combining both functions of disaster prevention and environmental improvement.

In both surveys at the port of Wajima and the port of Shimonoseki where EAC is actually used, it was confirmed that EAC becomes favorable foundation for living organisms, compared to the Ord-con. By using EAC for the submerged part of all concrete structures which are indispensable for disaster prevention such as breakwaters and revetment, EAC can achieve both "disaster prevention" and "environment conservation", further improvement of fishing ground productivity is expected.

Based on these success examples, SPF greatly contributes to Goal 13 of "Sustainable Development Goals (SDGs)" set by the United Nation in terms of disaster prevention by strengthen the resiliency against disasters such as storm surge accompanying climate change and tsunamis caused by earthquakes. At the same time, SPF also contributes to Goal 14 in terms of environmental conservation by creating the growth and habitats for living organisms to improve the fishing ground productivity. Likewise, EAC has already been certified by Japan Science and Technology Agency (JST) as an example for achieving SDGs in Japan and contributed to SDGs in terms of environmental conservation. Hereafter, it is hoped that these technologies will be used around the world beyond Japan and help to an achievement of both “disaster prevention” and “environmental conservation”, and realization of the sustainable society.

ACKOWLEGEMENT

I would like to express appreciation to director of Nikken Kogaku Co.,Ltd., Mr. Hisao OUCHI for his guidance and advice.

I would also like to appreciate to Ms. Yuka KIZAKI in International Division for her big support in writing this paper.

REFERENCES

Hiraishi, T., Mase, H., Kawata, T., Yukimoto, T., Tokunaga, S. and Matsushita, H.(2011). Experimental Study on Counter-weight Blocks for Breakwater Stability, Proc. of the 21th Int. Ocean and Polar Eng. Conf., ISOPE-2011, Volume 3, pp.1146-1151.

Matsushita, H.(2013). Breakwater Reinforcement Method against Large Tsunami, PIANC De Paepe-Willems Award, PIANC Yearbook 2013, pp.111-130.

Matsushita, H., Hiraishi, T., Mase, H. and Kishimoto, O.(2014). Performance evaluation of counter-weight block against accidental wave and tsunami actions, Proc. Civil Eng. In the Ocean, JSCE 70, No.2, 468-473 (in Japanese).

Nishimura, H., Yamanaka, R., Sato, K., Tara, C., Nakanishi, T. and Kozuki, Y.(2014). Material performance of "Enviromentally Active Concrete" containing amino acids in water, Pianc World Congress.

Sato, K., Kozuki, Y., Nishimura, H., Tara, C., Nakanishi, T., Nakata, H., Yamaguchi, N., Nogami, F., Yamanaka, R., Tabohashi, T. and Kobayashi, M.(2011). Characteristic of Periphytic Algae on Concrete Contained Amino Acid, 9th Int. Conf. on The Environmental Management of Enclosed Coastal Seas.

TECHNICAL STANDARDS AND COMMENTARIES FOR PORT AND HARBOUR FACILITIES IN JAPAN(2018), The Overseas Coastal Area Deveropment Institute of Japan (in print).

Seto Inland Sea Environmental Protection Basic Plan(2015), Ministory of the Environment.

Book of Japan's Practices for SDGs – Creating Shared Value by STI, Business and Social Innovation – (Preliminary Edition, Fall 2017), pp.36, Japan Science and Technology Agency

EXPLORING POTENTIAL CLIMATE CHANGE IMPACTS AND ADAPTATION STRATEGIES FOR SEAPORT OPERABILITY

Judith K. Mol[1,*], Wiebe P. de Boer[2,3], Tiedo Vellinga[2], Jill H. Slinger[2], Victor Beumer[3]

Keywords: climate change downscaling, port operability, adaptation strategies

ABSTRACT

As seaports are located within the coastal zone, they are susceptible to climate change impacts such as changing wave conditions and sea level rise. To secure the operability of seaports under these changing conditions, a deeper understanding of potential local-scale climate change impacts is needed to explore suitable adaptation strategies. Previous studies focused on climate change impacts on seaports at a regional scale. Consequently, there remains a lack of understanding of these impacts on individual seaports and how these seaports can accommodate or adapt to these impacts. This study provides a conceptual framework for (i) quantifying risks for port operability and (ii) exploring adaptation strategies. The framework is tested on a case study of the Port of IJmuiden in the Netherlands. The study demonstrates that in the absence of adaptation measures, climate change may result in significant risks for the operability of this port. While the framework is tested on a single case study site, it is believed to be a promising tool for exploring climate change risks and adaptation strategies for seaports worldwide.

1. INTRODUCTION

Ports are vital within the global economy due to their crucial role in the globalized trading system. Because of an expected increase in the world's merchandise seaborne trade, it is important their role will be maintained in the future (UNCTAD, 2011). The performance of ports depends on the level of operability in terms of the amount of uptime, defined as the extent to which it is possible to continue operations such as berthing, mooring and navigating.

Ports are often located in coastal zones due to strategic considerations, to optimize the accessibility between land and water. Consequently, seaport operations are exposed to specific physical conditions such as storm surges, storm waves and sea water levels. Therefore, seaports are vulnerable to climate change impacts on these conditions. For example, sea level rise causing increased water levels may result in flooding of quay infrastructure and subsequently hinder quay wall operations.

Following a worldwide survey amongst 93 seaports, port authorities are concerned about sea level rise and increased storminess, causing flooding of infrastructure and impacting port operations (Becker et al, 2011). To secure seaport operability under these changing conditions, an understanding of potential climate change effects and possibilities for adaptation is needed. Most port authorities feel uninformed about potential climate change impacts and consider climate change as a topic they should know more about (Becker et al., 2011). The United Nations considers assessing climate change impacts and adaptation options for ports to build their resilience as an urgent imperative (UNCTAD, 2011).

IPCC suggests approaches for assessing climate change impacts on a conceptual level (Lal et al., 2012). Other studies focus on climate change impacts at the scale of a sea basin, for example, the North Sea (Grabemann et al., 2014; Groll et al., 2013). A recent study provides a framework for quantifying local scale climate change impacts on coasts (Ranasinghe, 2016). However, the amount of publications on assessing climate change impacts on individual seaports is limited (Becker et al. 2011; McEvoy et al., 2013). Therefore, there is a lack of understanding of how ports can accommodate or adapt to these impacts.

---

[1] Royal HaskoningDHV, Amersfoort, THE NETHERLANDS
[2] Delft University of Technology, Delft, THE NETHERLANDS
[3] Deltares, Delft, THE NETHERLANDS
* Corresponding author. Email address: judith.mol@rhdhv.com

In order to improve the understanding of this topic, this paper provides a conceptual framework to assess climate change impacts on port operability and to explore potential adaptation options for individual seaports. The presented framework combines and builds upon existing approaches to assess climate change impacts (Grabemann et al., 2014; Groll et al., 2013; Lal et al., 2012; Ranasinghe, 2016) and is tested on a case study of the Port of IJmuiden in the Netherlands.

Chapter 2 reviews existing assessment approaches and introduces the developed framework. Application of the framework to the case study of the Port of IJmuiden is provided in Chapter 3. Chapter 4 discusses the relevance and suitability of the use of the framework to seaports worldwide and Chapter 5 provides the main conclusions of the study.

2. METHODOLOGY

Section 2.2 presents the framework for quantification of climate change risks for port operability and for exploring adaptation options. It combines and builds upon existing approaches resulting from a literature review described in section 2.1.

2.1 Review of existing approaches for climate change impact assessments

The approaches for assessing climate change impacts in previous studies are conceptual (Lal et al, 2012), but also provide concrete steps for quantification of impacts at regional scale (Grabemann et al., 2014; Groll et al., 2013) and local scale (Ranasinghe, 2016).

IPCC suggests two conceptual approaches. A top-down approach (Figure 1, left part) focuses initially on downscaling of global climate change impacts by modeling, to assess relevant impacts (Kwadijk et al, 2010; Ranger et al., 2010). These results provide the base for designing relevant adaptation options. The second is a bottom-up approach (Figure 1, right part) which suggests studying the context and its vulnerabilities first, to subsequently define the focus for identifying adaptation options. Studies recognize that both approaches are complementary and combining them can be beneficial for the result (Lal et al., 2012).

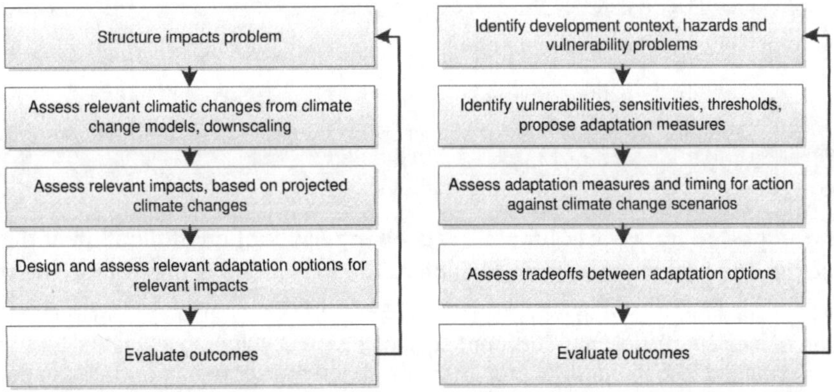

Figure 1: Existing approaches for assessing climate change impacts and for identifying adaptation options. Figures adopted from Lal et al. (2012). Left: the top-down approach, "Climate models, scenarios, impacts-first". Right: the bottom-up approach, "Vulnerability, threshold-first.

Downscaling global climate change predictions to the local scale is crucial for impact assessments of individual ports. Several studies provide concrete steps for downscaling global climate change impacts to local scale physical conditions (Grabemann et al., 2014; Groll et al., 2013; Ranasinghe, 2016). Following these approaches, a first step is to select global greenhouse gas (GHG) emission scenarios as input for Global Climate Models (GCM's, e.g. models such as HADCM and GISS). GCM output consists of time series of various climate variables, such as surface temperature, wind and precipitation. As the resolution of GCM output is too coarse for a local scale climate change impact assessment, a next step is to further downscale the result to a regional or local scale. This comprises the use of the coarse grid output of a GCM (typically in the order of 200 x 200 km) as boundary conditions of a model with a finer grid, such as a Regional Climate Model (RCM, e.g. models like CRCM and CCAM with a typical output in the order of 50 x 50 km). In order to obtain corresponding

regional scale climate change predictions on hydraulic conditions which are directly affecting port operations, such as water level or wave conditions, regional forcing models (e.g. WAM) are applied with the RCM output as boundary conditions. Ranasinghe (2016) suggests an additional step for downscaling regional scale impacts to the local scale on coasts, by using the output of the RCM and forcing model as boundary conditions of local scale coastal impact models.
Noted is that instead of dynamic downscaling, statistical downscaling is possible in areas in which data is abundantly available. Furthermore, bias correction of the results (e.g. of the RCM) is conducted to improve the accuracy of the output, by comparing obtained time series of physical conditions with field measurements and by applying correction techniques as required. Also, a range of scenarios is considered, for incorporation of uncertainties due to assumptions of global GHG scenarios. Finally, Ranasinghe (2016) suggests using an ensemble modeling approach to include modeling uncertainties, by using multiple global- and regional climate and forcing models.

2.2 Suggested conceptual framework for climate change impact assessments on port operability

The framework suggested for quantification of climate change risks for port operability and for exploring adaptation strategies is based on specific elements of the existing approaches presented in section 2.1.

Figure 2 displays the framework with its main steps: (1) analyzing the port system and selecting its valuable and vulnerable port operations and related infrastructural assets, (2) selecting global climate change scenarios and downscaling these to the local scale to assess the impacts within the port, (3) quantifying port risks and investigating their relevance and (4) exploring adaptation options for the port.

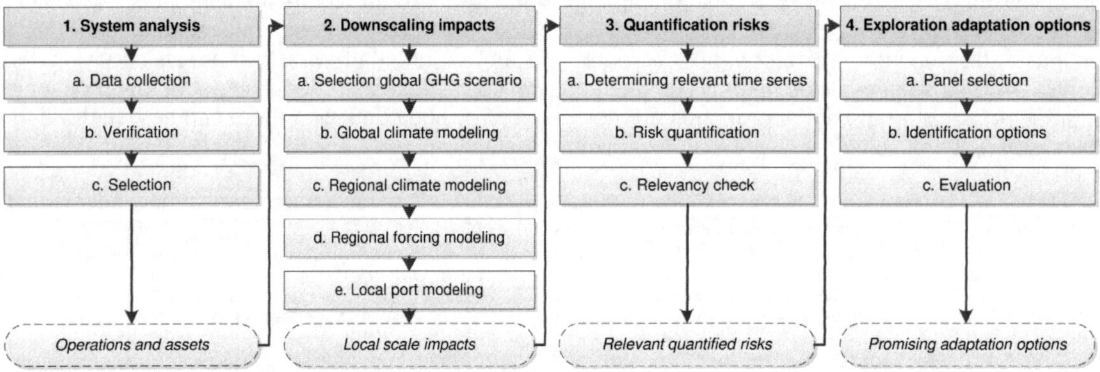

Figure 2: Suggested framework for a climate change impact quantification and adaptation study for ports. The box at the bottom of each of the four steps is describing the main output.

The framework combines the conceptual top-down and bottom-up approach of IPCC (Kwadijk et al, 2010; Ranger et al., 2010). Initially, the context including its vulnerable assets and thresholds are identified according to the bottom-up approach (step 1) and subsequently climate change impacts are assessed by means of downscaling, which is in line with to the top-down approach (step 2). Eventually, exploration of adaptation options (step 4) is following from both IPCC approaches. Specifications for downscaling impacts to the regional scale (step 2a to 2d) are adopted from the existing approaches presented in section 2.1 (Grabemann et al., 2014; Groll et al., 2013; Ranasinghe, 2016). Local port modeling (step 2e) in order to determine impacts within seaports is in line with the final step of the framework of Ranasinghe (2016) in order to determine local scale impacts.

Step 1 of the framework is a system analysis, consisting of selecting port operations and infrastructural assets that are valuable to stakeholders and that may be vulnerable to climate change impacts. These operations and assets form the focus of the subsequent climate change risk assessment. In addition, a focus must be set on specific climate change hazards, such as sea level rise, wind, storm surge or waves. For the specific case study of IJmuiden, climate change impacts on sea levels, storm surge and storm waves are considered.
An inventory is made (step 1a) of (i) operations and infrastructural assets, (ii) physical conditions and (iii) port operability thresholds values: values of physical conditions (e.g. wave height, water level or wind speed) and related variables (e.g. overtopping or flooding discharges) under which a specific port operation cannot take place and which is causing port downtime.

Stakeholders are identified, based on their potential vulnerability to climate change effects and their influence in the area. These actors are contacted in order to verify the collected data (step 1b) and to make a selection of valuable and vulnerable port operations and infrastructural assets (step 1c).

Subsequently, global climate change impacts are downscaled and assessed on the local scale of the port (step 2). Specific steps for determining these global impacts and for downscaling to the regional scale are adopted from the approaches presented in 2.1 (Grabemann et al., 2014; Groll et al., 2013; Ranasinghe, 2016). An ensemble scenario- and modeling approach is suggested to obtain insight in GHG scenario and modeling uncertainties. However, a tradeoff should be considered between computational model effort and the accuracy of the results.
Firstly, global climate change impacts are determined by means of a GCM with the input of a global GHG emission scenario (step 2a and b). The coarse grid GCM output presents a set of time series of climate variables (e.g. air pressure, wind, precipitation) for the present and future time slice. Subsequently (step 2c), regional impacts are determined by dynamic downscaling the GCM results, by using the coarse grid GCM output as boundary conditions of a RCM with a finer grid. To obtain corresponding predictions for hydraulic conditions directly impacting port operations (e.g. water levels, storm surge and waves), the RCM output is used as boundary conditions of a regional forcing model (step 2d).
Eventually (step 2e), local projections within seaports at an appropriate scale (depending on the port, but typically in the order of 10-100 m) are determined by using the RCM output (e.g. wind) and output of regional forcing models (e.g. water levels, storm surge and wave characteristics) as boundary conditions of an appropriate local validated model (e.g. Delft3D). Alternatively, GCM output is used as boundary conditions of the local model directly, but this might lead to a less accurate result.

In order to quantify port risks and to investigate their relevance (step 3), the following three steps are required.
Initially (step 3a), additional data is obtained based on analysis of local scale physical impacts following from step 2 (e.g. time series of mean sea levels, storm surge and wave characteristics). Gathered data are time series of variables related to the defined port operability thresholds (e.g. data on overtopping discharges), which are assessed by means of rules of thumbs with input of the local scale impacts. For example, suggested rules of thumbs for determining overtopping discharges are theories by Allsop (1995) and Franco (1994), applicable depending on the relation between the water depth and significant wave height at a specific moment (van der Meer & Bruce, 2014).
Then, climate change risks are quantified as expected changes of frequencies of port downtime (e.g. in terms of expected amount of days per year) in the future time slice compared to the present time slice (step 3b). Port downtime is defined as the time in which it is not possible to continue port operations, due to exceedance of a port operability threshold value by one of the relevant physical conditions or related variables (e.g. wave heights exceeding the wave height threshold value for navigational operations). By including all relevant physical conditions in the same analysis of a specific port operation, correlations between different types of physical conditions leading to port downtime are taken into account (e.g. correlated high water levels and wave heights during a storm event causing a flooded quay wall and therefore downtime). In case of a lack of data points above the port operability threshold values (e.g. water level heights which are lower than threshold values for flooding in all data points of the time series), return periods are obtained by means of a Peak over Threshold analysis, based on Extreme Value Theory (Caires & van Os, 2012).
Eventually, the relevancy of the risks defined is checked by the input of stakeholders (step 3c).

Due to the uniqueness of most ports and their environment, promising adaptation strategies might differ for each case. In order to explore adaptation options (step 4) and to obtain a Taylor-made solution for a port, a panel is composed (step 4a), consisting of a combination of stakeholders and experts with backgrounds in seaports, policy, management and other relevant subjects to the case, such as specific physical conditions (e.g. hydraulics). Initially, a divergent brainstorm session is conducted (step 4b) in the categories (i) operational: logistic and technological solutions, (ii) institutional: economic, legislative and political options, (iii) social: options in which the influence on the behavior of actors is considered, (iv) grey: physical engineering solutions within the port area and (v) green: win-win solutions which are also benefitting the natural environment. These categories are based on suggestions by IPCC (Wong et al., 2014). In a subsequent convergent part (step 4c), promising alternatives are selected by means of voting and a panel discussion.

3. APPLICATION OF THE FRAMEWORK TO THE PORT OF IJMUIDEN

A case study is conducted to illustrate the application of the framework. Selected is the Port of IJmuiden in the Netherlands, approximately 20 kilometers west of Amsterdam. An overview of the port and the focus locations for this study is displayed in Figure 3.

Figure 3: Study locations within the Port of IJmuiden notified in red with (1), (2) and (3). The photo is adopted of the Netherlands Space Office (2015). The location of the port within the Netherlands is displayed in black on the map of the country.

In section 3.1, the results of the system analysis of the Port of IJmuiden are presented. Section 3.2 discusses the downscaled impacts and section 3.3 provides the climate change risks for port operability in terms of changed frequencies of downtime. In section 3.4, promising adaptation options for the case study of the Port of IJmuiden are introduced.

3.1 Port operations and infrastructural assets

Selected port operations and assets in the Port of IJmuiden are (i) navigational activities (Figure 3, location 1), (ii) berthing and mooring of vessels (Figure 3, location 2) and (iii) quay wall operations (Figure 3, location 3). These operations and assets are verified to be valuable to stakeholders. Climate change impacts on sea levels, storm surge, wind waves and overtopping discharges are assessed since the selected assets and operations are sensitive to these physical conditions or related variables.

For each of the types of operations, threshold values for port operability and their specifications are displayed in Table 1.

Table 1: Specifications of port operability thresholds in terms of significant wave height (H_s) and overtopping discharge (q) for the selected operations in the Port of IJmuiden.

Type of operations	Vessel specification	Threshold values for port operability	Criterion angle of incidence waves	Reference
Navigational activities	Tugboat Assistance	**H_s (m) = 2.5**		Svitzer Amsterdam (2016)
Berthing and Mooring	Bulk	**H_s (m) = 1.0**	θ (degrees) = 0-45 (head or stern)	Ligteringen & Velsink (2012)
	General cargo	**H_s (m) = 1.0**	θ (degrees) = 0-45 (head or stern)	
	Offshore and Wind farm	**H_s (m) = 0.8**	θ (degrees) = 45-90 (beam)	

	Fishing	**H_s (m) = 0.8**	θ (degrees) = 45-90 (beam)	
	Transportation and Roro	**H_s (m) = 0.5**	θ (degrees) = 45-90 (beam)	
Quay wall operations		**q (m^3/m/s) = 0.05**		Vrijling et al. (2011)

3.2 Downscaled climate change impacts within the port

As input of the modeling assessment, the Representative Carbon Pathway (RCP) scenario (range) 4.5-6.0 is assumed, an emission scenario range which is in line with the amount of global gasses emitted in case of a climate policy being successfully implemented (Moss et al., 2010). In this study, this scenario will be referred to as the 'moderate emission scenario'. In addition to that, the RCP scenario 8.5 is considered, a higher emission scenario which is referred to as the 'high emission scenario'.
Global and regional climate change predictions for the significant wave height, storm surge height and sea level rise for each of the assumed scenarios are following from an ensemble modeling approach. Previously to this study, numerous other studies are published in which predictions for the region of the Port of IJmuiden (especially the North Sea area) are obtained. This study builds upon these results by taking representative predictions for each of the assumed global GHG scenarios. These predictions for regional scale impacts are downscaled to the local scale, by using them as boundary conditions of the local validated SWAN model for the Port of IJmuiden (Booij et al., 1996).

3.3 Risks port downtime

The resulting quantified risks for the selected operations (i) navigational activities (location 1), (ii) berthing and mooring of ships (location 2) and (iii) quay wall operations (location 3) are displayed in Table 2.

Table 2: Expected frequencies port downtime per time slice due to hinder of selected operations.

Location	Operation	Reference period Time slice 1979-2001	Moderate emission scenario RCP 4.5 – 6.0 Time slice 2070-2100	High emission scenario RCP 8.5 Time slice 2070-2100
1	**Navigating**	*5 days/year*	*7 days/year*	*10 days/year*
2	**Berthing / Mooring**	*1 days/year*	*1.15 days/year*	*2 days/year*
3	**Quay wall operations**	*1 day / few hundred years*	*1 day / few years*	*1 day / 1-2 months*

The risks identified are relevant since the port is operating 24 hours per day and continuity in the operations is required.

3.4 Adaptation options

Based on the expert voting and panel discussion, the following conceptual adaptation options are identified as promising in dealing with sea level rise: (i) increasing quay heights and (ii) construction of a retention basin in combination with a drainage system for quay walls. The first option (i) is considered promising due to its effectiveness in reducing the risk of flooding of quay walls and due to expected low investment costs. The second option (ii) is besides being regarded effective also beneficial to the natural- and social environment. To deal with increased downtime due to changed storminess, the following options are identified as promising: (i) a multi-purpose land reclamation offshore of the port to provide a shelter zone for high waves and (ii) the application of navigable, wave-absorbing vegetation between breakwaters. The land reclamation (i) is stated to be a promising option benefitting the shipping- and airport industry, whilst being risk reductive. Option (ii) is positively influencing the natural environment whilst reducing wave energy. Further studies on the feasibility and effectiveness of all options are still required.

4. DISCUSSION

Regional predictions for sea level rise and increased storminess are more extreme in other regions than the area of the Port of IJmuiden. For example, wave heights are projected to increase more rapidly due to climate change at the German and Norwegian border of the North Sea, in southern and eastern parts of Australia, south-east Asia, west- and east Africa, South-America and the Caribbean (Hemer et al., 2012). This implies that climate change risks for port operability may be even larger for seaports located in these areas, and calls for attention on this issue worldwide.

It is believed that the conceptual framework presented in this paper is useful for assessing climate change impacts on seaport operability in terms of downtime, due to wind waves affecting vessel operations and changed sea levels, storm surge and wave overtopping discharges causing delay of quay operations, as shown in the case study of the Port of IJmuiden. The framework can also be applied to seaports elsewhere for which these hazards are relevant since the suggested sub steps, models and methods are applicable more generically than just for the case study of the Port of IJmuiden. Assumptions made for the case study can be suitable for other ports as well, specifically regarding the global GHG scenarios and port operability threshold values for shipping and quay wall operations (e.g. assumptions regarding wave height conditions in which a ship can navigate sufficiently and safely).
The applicability of the framework for assessing other seaports depends on the availability of data of the port environment and on the specific hazards selected for the assessment. Data of port operations and physical assets to be analyzed (e.g. quay heights and types of vessels operating in the port) and other environmental specifications (e.g. bathymetry) are required. In case of availability of regional climate change predictions, such as for the Port of IJmuiden, at least regional measurements are needed (e.g. water levels, wave and wind characteristics) for downscaling the regional predictions to the local scale. Although the analysis of climate change induced changes in sea levels, storm surge, storm waves and subsequent wave overtopping discharges impacting port operations is illustrated by means of the case study of the Port of IJmuiden, the assessment of other hazards (e.g. changes in precipitation and temperature) and indirect impacts of sea level rise and increased storminess (e.g. changes in currents and sediment transport) is not. However, the framework can be extended by assessing (other) seaports to a larger range of hazards or indirect impacts, by applying appropriate models and vulnerability threshold values related to the relevant hazards and operations selected.

5. CONCLUSIONS

Seaports are located in the coastal zone and hence their operations are exposed to specific physical conditions (e.g. storm surges, storm waves and varying water levels), and therefore they are susceptible to climate change impacts on these conditions. For seaports to still be able to operate under these changing conditions, it is crucial to understand potential climate change impacts and corresponding adaptation options.

This paper provides a conceptual framework to assess climate change impacts on seaport operability and to explore potential adaptation options. An initial step of the framework is (1) analyzing the port system and selecting its valuable and vulnerable port operations and related infrastructural assets, followed by a step (2) to select global climate change scenarios and to downscale these to the local scale to assess the impacts within the port. In a subsequent step (3), port risks are quantified and their relevance is investigated. In a last step (4), options for adaptation of the seaport are explored.

The framework is applied to the Port of IJmuiden illustratively, for which the impacts of climate change on sea levels, storm surge, wind waves and overtopping discharges on the delay of vessel- and quay operations are assessed, according to the first step. The results of the second and third step show that, without adaptation measures, climate change holds significant risks for port operability in the Port of IJmuiden. For each of the identified risks, promising conceptual adaptation options are identified following from the fourth step of the framework, although further studies on their feasibility and effectiveness are still required.

It is believed that the conceptual framework presented in this paper is useful for assessing climate change impacts on seaport operability in terms of downtime, due to wind waves affecting vessel operations and flooding of quays as a consequence of changed sea levels, storm surge and wave overtopping discharges. The suitability of the framework for assessing other seaports depends on the availability of data for a specific seaport environment. The framework can be extended by assessing (other) seaports to a larger range of hazards or indirect impacts (e.g. precipitation, currents and sediment transport). Therefore, the framework is considered promising for exploring climate change risks and adaptation options for seaports worldwide.

ACKNOWLEDGEMENTS

This work is funded by Deltares, Royal HaskoningDHV and Delft University of Technology, which is highly appreciated by the authors. The authors would like to thank Rijkswaterstaat, the Port of IJmuiden, Svitzer Amsterdam and the Netherlands Space Office for their contributions to the study. Furthermore, thanks to Arie Mol, Gerben Hagenaars and others for reviewing and for providing useful suggestions which have improved this article.

REFERENCES

Allsop, N. W. H., Besley, P., & Madurini, L. (1995). Overtopping performance of vertical walls and composite breakwaters, seawalls and low reflection alternatives. Final Rep. of Monolithic Coastal Structures Project, Univ. of Hannover, Hannover, Germany.

Becker, A., Inoue, S., & Schwegler, B. (2011). Considering Climate Change: A Survey of Global Seaport Administrators. Stanford University.

Booij, N., Holthuijsen, L. H., & Ris, R. C. (1996). The "SWAN" wave model for shallow water. 25th International Conference of Coastal Engineering. Florida.

Caires, S., & van Os, J. (2012). ORCA Guidelines. Delft: Deltares.

Franco, L., de Gerloni, M., & Van der Meer, J. W. (1994). Wave overtopping on vertical and composite breakwaters. Proc., 24th Int. Conf. on Coastal Engineering, ASCE, Reston, VA, 103021045.

Grabemann, I., Groll, N., Weisse, R., & Möller, J. (2014). Climate change impact on North Sea wave conditions: a consistent analysis of ten projections. Ocean Dynamics, 255-267.

Groll, N., Grabemann, I., & Gaslikova, L. (2013). North Sea wave conditions: an analysis of four transient future climate realizations. Ocean Dynamics, 1-12.

Hemer, M.A., Fan, Y., Mori, N., Semedo, A., & Wang, X.L. (2013). Projected changes in wave climate from a multi-model ensemble. Nature Climate Change, 471-476.

Kwadijk, J. C., Haasnoot, M., Mulder, J. P., Jeuken, A. B., van der Krogt, R. A., van Oostrom, N. G. C., Schelfhout, H. A., van Velzen, E. H., van Waveren, H., & de Wite, M. J. M. (2010). Using adaptation tipping points to prepare for climate change and sea level rise, a case study for in the Netherlands. Wiley Interdisciplinary Reviews: Climate Change 1, 729-740.

Lal, P. N., Mitchell, T., Ahuld, H., Mechler, R., Miyan, A., Romano, L. E., & Zakaria, S. (2012). National systems for managing the risks from climate extremes and disasters. A Special Report of Working Groups 1 and 2 of the Intergovernmental Panel on Climate Change (IPCC). In C. B. Field, V. Barros, T. F. Stocker, D. Qin, D. J. Dokken, K. L. Ebi, M. D. Mastrandrea, K. J. Mach, G. K. Plattner, S. K. Allen, M. Tignor, P. M. Midgley (eds), Managing the Risks of Extreme Events and Disasters to Advance Climate Change Adaptation, 339-392. Cambridge, UK, and New York, NY, USA: Cambridge University Press.

Ligteringen, H., & Velsink, H. (2012). Ports and Terminals. Delft: VSSD.

Moss, R. H., Edmonds, J. A., Hibbard, K. A., Manning, M. R., Rose, S. K., van Vuuren, D. P., Carter, T. R., Emori, S., Kainuma, M., Kram, T., Meehl, G. A., Mitchell, J. F. B., Nakicenovic, N., Riahi, K., Smith, S. J. Stouffer, R. J. Thomson, A. M., Weyant, J. P., & Wilbanks, T. J. (2010). The next generation of scenarios for climate change research and assessment. Nature, 747-756.

Netherlands Space Office (2015, October). Airbus Spot-6 image.

Ranasinghe, R. (2016). Assessing climate change impacts on open sandy coasts: A review. Earth-Science Reviews, 320-332.

Ranger, N., Millner, A., Dietz, S., Fankhauser, S., Lopez, A., & Ruta, G. (2010). Adaptation in the UK: a Decision Making Process. London, UK: Policy Brief, Grantham Research Institute on Climate Change and the Environment and Centre for Climate change Economics and Policy.

Svitzer Amsterdam. (2016, November). Questionaire. J. K. Mol, Interviewer.

UNCTAD (Composer). (2011). Ad Hol Expert Meeting on Climate Change Impacts and Adaptation. Geneva, Swiss.

Van der Meer, J., & Bruce, T. (2014). New physical insights and design formulae on wave overtopping at sloping and vertical structures. Journal of Waterway, Port, Coastal and Ocean Engineering.

Vrijling, J. K., Kuijper, H. K., van Baars, S., Molenaar, W. F., van der Hoog, C., Hofschreuder, B., & Voorendt, M. Z. (2011). Manual Hydraulic Structures. Delft: Delft University of Technology.

Wong, P. P., Losada, I. J., Gattuso, J. P., Hinkel, J., Khattabi, A., Mclnnes, K. L., Saito, Y., & Sallenger, A. (2014). Coastal Systems and low-lying areas. In C. B. Field, V. R. Barros, D. J. Dokken, K. J. Mach, M. D. Mastrandrea, T. E. Bilir, M. Chatterjee, K. L. Ebi, Y. O. Estrada, R. C. Genova, B. Girma, E. S. Kissel, A. N. Levy, S. MacCracken, P. R. Mastrandrea, L. L. White, Climate Change 2014: Impacts, Adaptation, and Vulnerability. Part A: Global and Sectoral Aspects. Contribution of Working Group 2 to the Fifth Assessment Report of the Intergovernmental Panel on Climate Change, 361-409. Cambridge, United Kingdom and New York: Cambridge University Press.

The Panama Canal Water Filtration Plants. State of the art of the early 20th century and the early 21th century. Key elements for the Panama Canal Operations

By

Luz Meneses[1], Eric Vicente Rodriguez[2], Luis Santanach Bernal[3]

PANAMA CANAL AUTHORITY

WATER DISTRIBUTION PACIFIC SECTION

ABSTRACT

The construction of the Panama Canal by the United States of America started in 1904, right after Panama achieved its independence from Colombia and became a Republic on November 3rd 1903. The initial stages demanded the establishment, among other important issues, of a reliable drinking water system for the work force of the Isthmian Canal Commission (ICC) and the population of the adjacent cities of Panama and Colon. By November 1904 there were already 3,500 U.S. citizens hired to start the works for the new Canal.

Before and during the Panama Canal Construction, other sources of water supply were used successfully. The rivers running south toward the Pacific Ocean, Grande and Cocoli, were the sources for water consumption, coping with the demand and quality of water required.

Years later The United States Of America approved the construction of two Water Treatment Plants; Mount Hope in Colon (1914) and Miraflores Filtration Plant in Panama (1915).

On December 31 1999 the Republic of Panama took over the operation of the Canal and the new Panama Canal Authority (ACP), replaced the U. S. Federal Agency Panama Canal Commission (PCC). Under the ACP the new Water Treatment Plant "Mendoza" was built in 2010.

Indeed, having a Water Filtration Plant is a privilege that very seldom someone will find in marine facilities such as ports, navigation channels, marinas, ferry structures. But having three; people will be skeptical of this statement.

The Panama Canal is the exception to the rule. These plants provide drinking water for the cities of Panama, Colon, Arraijan, Veracruz, Chorrera; and the Canal employees including Canal operations. And they have been doing it since the beginning of the opening of the Panama Canal.

The water quality has prevailed through time regardless the continues changes on the regulations and also keeping the population from water borne diseases such as cholera whose outbreaks hit South America and The Caribbean in different waves during the 19th and 20th century. This Water Filtration Plants are part of the continued operation of this ingenious and marvelous structure of the engineering known as the Panama Canal.

---

[1] Panama Canal Authority, Supervisor Miraflores Filtration Plant. Industrial Eng. lmeneses@pancanal.com

[2] Panama Canal Authority, Supervisor Mendoza Filtration Plant. Industrial Eng. EVRodirguez@pancanal.com

[3] Panama Canal Authority, Water Distr. Mgr. Pacific Branch. M. Eng. Civil Eng. lsantanach@pancanal.com

A new challenged arrived during the Canal Expansion Program, providing potable water to ACP workers during a nine (9) year–period. This challenge continues with the new Locks and other facilities. These benefits even though intangibles are embedded silently on the operation.

The paper gives a briefing description of Miraflores and Mendoza Filtration Plants Processes. It also shows some general information related to amount of citizens and workers during different decades; as well as the regulations that the plants have complied through time. General information of previous water supplies structures is included on the paper.

Human factor remains a critical element to assure the good operation of the plants, administrative and technical teams work together to make sure potable water complying with regulation requirements is always available for the consumers.

INTRODUCTION

The main purpose of any water filtration plant is to produce water that is safe for drinking. This means, water free of chemical and biological organisms that may cause harm to humans to the point of developing an epidemic.

In 1904 when the United States of America initiated the construction of the Panama Canal, the building of some facilities was required to accomplish the task. Sanitary facilities as well as water treatment facilities were critical to keep the workforce free from diseases.

During the works performance a labor force of approximately forty five thousand (45,000) arrived from the Caribbean islands. Panama City population at that time was about twenty five thousand (25,000).

The Panama Canal opened to the world on August 15 1914. Since then it has served the world thanks to its effectiveness and efficiency on the operation as well as having a professional well trained, and though very seldom mentioned, "Healthy" labor force.

Before 1914, during the construction period the priority besides building the canal was the safety and health of the Canal workers and their families. Drinking Water was part of the equation for a healthy population.

Despite the continuous changes on regulations and engineering state of the art on Water Works, Panama Canal Filtration Plants cope with these demands enforced by the local laws and engineering technology.

In addition to the objective presented on the abstract, the purpose of this paper is also to:

a. Present the features of two of the three plants belonging to the Panama Canal.
b. Present the consumption and demands required trough time.
c. Market the importance of the Water Filtration Plants as key elements for the health of the work force.
d. Present that despite the changes of regulations the Plants still cope with the new ones.

MOUNT HOPE FILTRATION PLANT (MHFP) is not part of this paper, however is important to highlight that MHFP was the first Plant built in 1914. Therefore, its influence on the construction and performance of MFP is significant since the data acquire in relation to the behavior of each one of the components in MHFP and lessons learned during the construction aided to predict the behavior of MFP, built a year later.

This latter comment represents the milestone on the performance of the plants since there was no existing data or information earlier recorded.

Potable water is also used for the Chilled Water Plant, and Canal operation Fire Fighting Stations.

Now External demand is 95% of the total potable water produced by the Panama Canal. ACP sells potable water to the Panamanian Institute of Potable and Sewerage Waters (IDAAN), who is responsible for the distribution system to the final customer.

The water is provided to Panama City, Colon, Albrook Los Rios, Cardenas, Corozal, Paraiso, Pedro Miguel, Gamboa, Panama Police Academy, Gamboa Penitentiary, The new Hotel Resorts, Arraijan , Howard and Chorrera.

DEMOGRAPHY AND WATER CONSUMPTION (GENERAL INFORMATION)

Starting in 1904 until the present time, population and clients of the Panama Canal Filtration Plants have changed. Nevertheless the goal to provide a high quality drinking water remains the same.

In early years of the Republic of Panama the population in the Capital Panama City was around 50,000 according to Canal Zone Public Works Division. Water consumption was 246 liters (65 gallons) per day per person or the equivalent to 12,301,000 liters (3,250,000 gallons approximately) per day of total consumption.

In 1910 Canal workforce was over 50,000 where 6,000 were U.S. citizen. The rest of the workforce came from the West Indies, Europe and other parts of the world (figure. 1). The consumption 13,248,000 liters (3,500,000 gallons) per day. In 1913 this consumption increased to 28,388,000 (7,500,000 gallons) per day. Overall population including Panama and Colon Cities total 100,000 residents.

Figure 1: Europeans workforce.

In 1920 Canal Zone population was 22,000 inhabitants. During this decade the plants provided services for the following clients.

1. Domestic and industrial consumption for the cities of Panama and Colon
2. Domestic and industrial consumption for the communities of the Panama Canal at Cristobal, Mount Hope, Gatun, Paraiso, Pedro Miguel, Balboa, Balboa Heights and Ancon.
3. Domestic and industrial consumption to the military and naval forces posted at the Canal Zone.
4. Potable water production demand covers the three locks along the Panama Canal as well as all Canal Operation Facilities and Administrative offices.
5. To the vessels making up the canal ports.

The total population was approximately 116,000 persons not including the ship crew and tourists.

In 1930 the Canal Zone population including civilians and military personal was estimated to be 22,858 persons. The consumption 29,145,000 liters (7,700,000 gallons) per day. Population in Panama and Colon Provinces were 180,000.

Table 1 shows the approximate census taken from 1911 to 1970. Panama District had a population of 386,000 in 1979. It is important to point out that not all inhabitants were connected to the distribution system of the Panama Canal.

POPULLATION IN THE PROVINCES OF PANAMA AND COLON 1911-1970		
YEAR	PANAMA	COLON
1911	61855	32092
1920	98035	58250
1930	114103	57161
1940	173328	79119
1950	248337	90144
1960	372393	105416
1970	576645	134280

Table 1: Panama and Colon Provinces population

By 1954 PCC force total 15,000 employees according to the Canal Zone Census.

Census taken from 1979 to 1998 showed that the workforce population varied between 9,000 and 10,000 employees including U.S Citizens, Panamanian Citizens and Third Nationalities. During that period the Plants provided also water to the U.S. Army, U.S. Navy and U.S. Air force Stations in Panama.

By the year 2018 about ten thousand (10,000) employees in the Panama Canal. Panama City communities connected to the old PCC water lines still receive potable water. About 200,000 residents live in those communities. This number does not count the local business. Ancon which is the former Canal Zone area and former U.S. Military installations is included in that number. According to the latest census projection the population reaches the 49,000 residents.

Despite human migration the distribution system has not changed much.

PREVIOUS STRUCTURES (EARLY 20th CENTURY)

Among ICC major concerns was supplying suitable drinking water for domestic and industrial uses during the early years of the construction of the Canal.

The first water supply constructed was the Rio Grande Reservoir south of Culebra (figures 2 & 3) located along the old Panama Railroad tracks. Another reservoir at Culebra Cut (Gaillard Cut) on the other side of the continental divided was used by U.S. Military forces. Its name Camacho (figure 4)

Americans took advantage of the reservoir built by the French Canal Company. They raised a 5.18 meter-high (17 feet) dam getting a spillway elevation from mean sea level of 71.6 meters (235 feet).

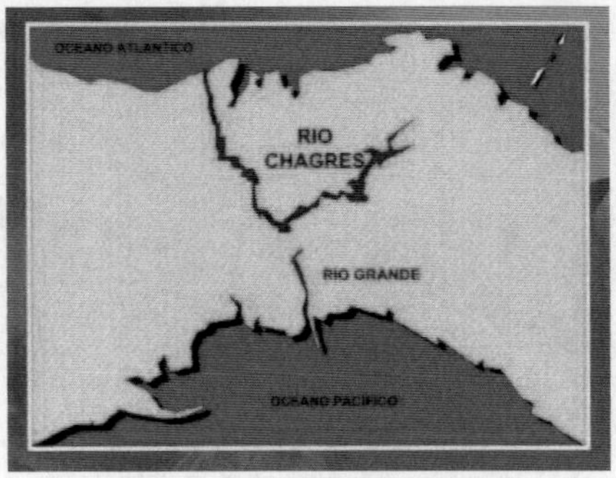

Figure 2: Grande River.

Figure 3: Grande River Reservoir (1904-1914)

The reservoir effective capacity was one billion one hundred ninety three million (1,193,000,000) liters or three hundred fifteen million (315,000,000) gallons. Water source was the Grande River flow and Panama tropical rain. To deliver water from Culebra to Ancon a 404.4 mm (16inch) cast iron (bell and spigot) pipe was installed along the 15.85 km (52,000ft) distance.

A 3,785,000 liters (1,000,000 gallons) capacity concrete reservoir was also built at Ancon. The system was capable to supply water due to gravity up to an elevation of 30.48 meters (100ft).

The pipe distribution was complemented with a 20-inch cast iron pipe (506 mm). This both pipes were connected together into a 24 inch pipe (607 mm) at the north end of Pedro Miguel locks west lane.

Before 1913, water supply was also provided, when required, by Cocoli Lake. The operation was executed using electric driven pumps. This water shed was temporarily used. The System supplied water at Panama City through filters.

Cocoli was later discontinued due to the flooding and creation of Miraflores Lake.

Figure 4: Camacho reservoir

THE NEW 1915 PROJECT. MIRAFLORES FILTRATION PLANT. (figure 5)

Figure 5: Miraflores Filtration Plant 2018.

Water quality and water demand called for the construction of a new plants. The United States Government approved the construction of Water Filtration Plants within the Canal Zone.

Upon approval, the decision was to build the new plant on the east bank over the hill near northeast of Miraflores Locks. The ground elevation was 35.36 meter (116 feet) above sea level and 18.59 meters (61 feet) above Miraflores Lake level. The plant must have the capability to provide low and high service through a single distribution system. The general specifications included (using original Unit system):

a. Maximum nominal capacity per acre: 125,000,000 gallons
b. Flow Rate per day: 12,000,000 gallons
c. Maximum capacity at the above rate: 15,000,000 gallons.

Once prepared the scope of work, the budget and the construction approval, Miraflores Water Filtration

Plant construction was successfully concluded on March 16 1915 at a cost of One Million Three Hundred Eighty Seven Thousands Four hundred Ninety U. S. Dollars with twenty two cents ($ 1,387,490.22). The plant previous features included:

a. An aeration system: When building Gatun Lake, at that time the largest manmade lake in the world, a lot of tropical forest as well as marine plants due to the swap areas were flooded producing hydrogen supplied gas due to the decomposing vegetable matter. Aeration eliminated odor as well as amounts of iron coming from Miraflores Lake, Chagres River, and Gaillard Cut, where the intakes were built after the dismantling of Miraflores intake.
b. Head house and Mixing Chambers: To inject Aluminum sulfate to the raw water.
c. Sedimentation Basin: Three sedimentation with the overall capacity of 1,161,290.32 liters (4,500,000 gallons)
d. Filter building: Fourteen (14) filters with a capacity of 157 liters/sec (2500 GPM) each.
e. Clear Well: Built with to two exists where water was transported by gravity
f. Pump Station (PS2 Miraflores)
g. Injection Chamber
h. Effluent controllers
i. Aluminum and hypochlorite mixing apparatus.
j. Hypochlorite of lime dosing apparatus

k. Pump Raw water intake (figure 6.): the first location was chosen to be built at the foothill of the plant over the Miraflores Lake. Nevertheless, due to the heavy concentration of the chlorine in the water the construction was abandoned. ICC decided to build the intake station on the bank of the Chagres Rivers north of Gamboa reach. This Intake station si still in operation. Distance to the Plant was approximately 16 Km (10 miles) (figure 3.). To transporting raw water a 17.3 km (56,762 linear feet), of cast iron pipe was installed.
l. Booster pump station at Balboa (U.S. Pump station 1)

The pipe diameter varied from 24 to 36 inches. The scope consisting on the pipe purchase, installation which included backfilling and excavation work had a cost of Three hundred Fifty Six Thousands Nine hundred Fifteen U.S. Dollars ($ 356,915)

Figure 6: Gamboa Intake Station At Chagres River

The total cost of wash water tank, aeration basin, the head house, the sedimentation basin, the filter building, pipe gallery, laboratory, quarters, clear-water basin, and the injection chamber cost was $558,168.

Distribution of the potable water consisted on three parallel lines 6,400 meters (21,000 ft.) in length each. Diameter varied in sizes between 759 mm (30 inches), 506 mm (20 inches) , and 404.8 mm (16 inches).

The three pump stations (Balboa, Miraflores, and Gamboa), cost Two Hundred Thousand Six Six hundred Fifty Four U.S Dollars with Fifty Seven cents ($200,654.57).

A high service reservoir was built at Ancon Hill, located at an elevation of 91.4 meters (300 feet) above sea level approximately with a capacity of 5,677,500 liters (1,500,000 gallons).

Currently there are eleven potable water reservoir tanks located strategically along the system with a maximum capacity of 51.15 ML (13.2 MG)

MFP PLANT PROCESS 1915 to PRESENT

Gamboa Intake

Actually Gamboa is a 4 -pump system where three have a flow rate of 6,400 GPM and one of 4,200 GPM. Since the limitation is the pipe capacity the combine between pumps can give ranges of maximum flow between 946 to 1,262 liters/sec (5,000 and 20,000 GPM)

Paraiso Intake (figure 7)

Paraiso possess five (5) low lift pumps and three (3) booster pumps. The combination between pumps due to pipe capacity range 479 to 2, 271 liters/sec (7,600 to 36,000 GPM).

Figure 7: Paraiso Intake Station

Main Plant (1914 to present)

Miraflores Filtration Plant is designed to process raw water with the conventional treatment method of the 1900s; the treatment is accomplished with chemical products for coagulation, flocculation, and odor and taste control and for disinfection.

At Miraflores Plant, the infrastructure has a gravity process starting with the aeration basin. Before entering the aeration, basin chlorine is injected.

On the head house and the mixing chambers, aluminum sulfate, and carbon activated are injected as well as polymer. Panama Canal plants are the only that inject fluorine.

The sedimentation basins have a capacity of 17 ML (4.5 M gallons). The filters pools are composed with gravel, sand and anthracite. Twenty filters form the filter pool. The plant was expanded in 1941 incrementing the number of filters and the Clear Well Capacity to 516,129 liters (2 MG)

The clear well expansion in 1941 is designed with five exits. Three run the water by gravity and the other two take the water to two pump stations located in Miraflores. The first pump station was built with the plant. It is called PS2. The second one built in their 2000s. This latter delivers potable water to west side communities such as Arraijan and Veracruz.

The plant sand filters, due to increase in potable water demand from these plants, were renovated in the late 1970´s, anthracite was incorporated over the sand, the filter layer design was resized and a surface wash water system was installed.

Treatment up to this time remains conventional on most of the stages, aluminum sulfate, liquid chlorine are the primary chemical products; activated carbon is used, fluoride addition is required by the Republic of Panama law, Miraflores uses polymers in different times of the year when process results show that its use is cost effective and efficient in the quality of the water in process.

Reservoirs are part of a distribution system that mostly has pumping stations, pipelines for filling tanks and transferring the water to the city of Panama at points designated as "delivery points" that also serve as the boundary for Canal responsibility which ends at these points.

The outside view of these centennial plants remind the typical architecture from the past, however inside them the new technology has been incorporated by means of monitoring instrument, remote control for pumps and valves, electronic devices and equipment, pressure and level sensors, all interconnected to a Supervisory and Data Acquisition System (SCADA) that provides real time information for water treatment operators.

The plant can stand up to 200 NTU at the raw station and delivered potable water with a NTU less than .35 Which is within the Local laws regulations. During the December 2010 heavy rain called the "Purisima" turbidity at the intake raised to 800 NTU. Despite that extraordinary event the Plant delivered the water within the numbers required. An schematic of the Pacific Distribution System is shown in figure 6

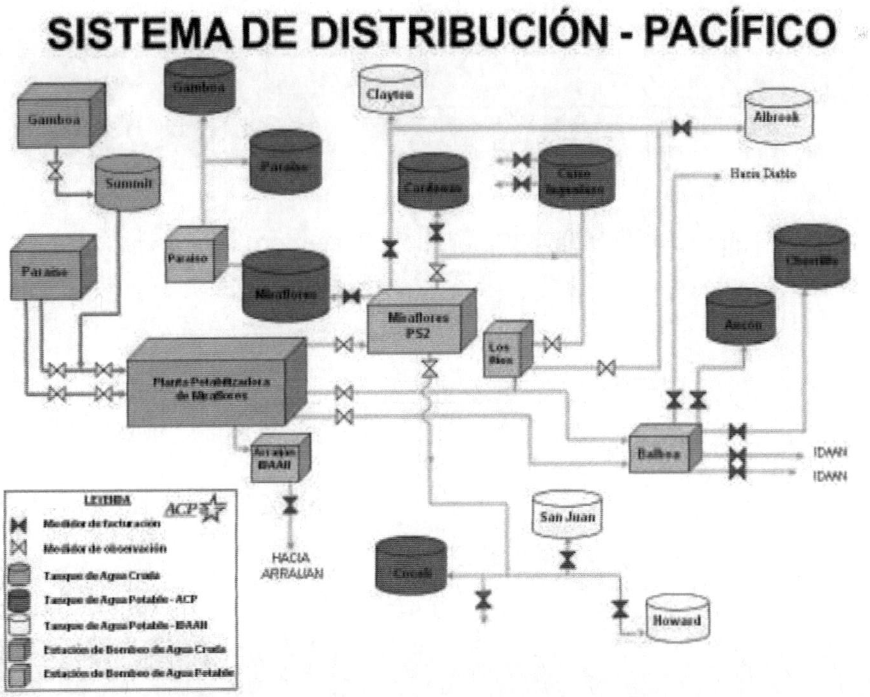

Figure 8: Pacific water distribution schematic

MENDOZA FILTRATION PLANT (MEFP) (21st century PROJECT)

MEFP, designed and built by Biwater International Limited, began operation in 2010, implementing the dissolved air flotation technology for the treatment of the Gatun lake water. Intake (figure 8) is located at La Represa town at a distance of 10km from the plant. The Plant has a capacity of 159 MLD (42 MGD) equivalent to 6624.47 cubic meters/hour for raw water. The potable water maximum production is 155 MLD (40 MGD) equivalent to 6,309 cubic meters/hour. Water flows into the rapid mixing basin where the dissolved air flotation process begins within a concrete structure that includes the areas for chemical deification, flocculation, filtration, potable water, filter backwash, water settling and sludge drying. Each compart dimension for the flocculation basin is 8m x 5 m with an average water depth of 4m. The volume for each water basin is 160 cubic meters.

MEFP possess five rapid mixing tanks, each with 1,329.92 m3/hour (5,408 GPM), five (5) flocculation lanes and twelve (12) dissolved air flotation (DAF) tanks .Figure 7 represents the schematic of water treatment process at Mendoza. Table 2. Shows the flow rate water treatment at MEFP.

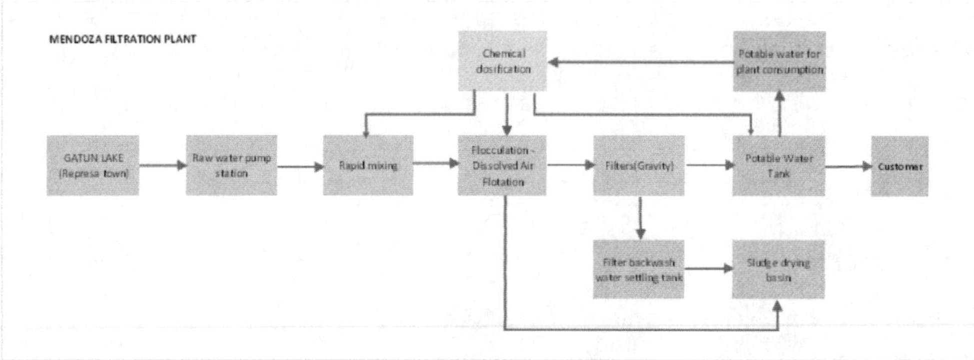

Figure 9: Mendoza Water Plant Treatment Process

FLOW RATE FOR WATER TREATMENT AT MENDOZA					
	Max	normal	MAX	MAX	MAX
DESCRIPTION	c.m./hr	cu.m/hr	GPH	GPM	GPD
Raw Water to Plant	6624	6624	1,747,757.26	29,129.29	41,946,174.14
To Inject Alminum Sufate	1.7	0.8	448.55	7.48	10,765.17
Cl precolation	23.2	23.2	6,121.37	102.02	146,912.93
Carbon	0.126	0.038	33.25	0.55	797.89
Sludge Removal	93.8	40.6	24,749.34	412.49	593,984.17
Entering the filters	6624	7286	1,747,757.26	29,129.29	41,946,174.14
Filter wash water	169.1	169.1	44,617.41	743.62	1,070,817.94
Filter water toward the tank	6386.4	6438.9	1,685,065.96	28,084.43	40,441,583.11

Table 2: Mendoza Water Treatment Flow Rate.

Figure 10: Mendoza Filtration Plant intake at Gatun Lake

GENERAL INFORMATION ABOUT THE LAWS AND REGULATIONS THAT REQUIRE COMPLIACE BY PANAMA CANAL WATER FILTRATION PLANTS

In 1914 some limited drinking water standards were implemented in U.S. cities.

In the 1940s federal drinking water standards were widely applied. In 1941 , the U.S. Department of The Treasury promulgated the country's first drinking water bacteriological standard, a maximum level of 2 coliforms per 100 mL.

Federal regulations and standards for water treatment in a more wide-ranging way were implemented in the 1970s.

In 1972, the Clean Water Act passed through Congress and became law , requiring industrial plants to proactively improve their waste procedures in order to limit the effect of contaminants on freshwater sources.

In 1974, the Safe Drinking Water Act (figure 11) was adopted by all 50 U.S. states for the regulation of public water systems. It identified contaminants that must be closely monitored. This reaction was due to the finding of organic chemicals on drinking water suspected of being carcinogens. Since then, the government, the public health community, and water utilities throughout US have worked together to safeguard the nation's drinking water supplies and to ensure that law protects public health in the best possible ways.

After year 2000 MFP and Mendoza Plant (2010) had to comply with Panama Ministerio de Salud (MINSA) ("Health Ministry") and the Autoridad de Servicios Publicos (ASEP) ("Public Services Authority"). Mendoza Filtration Plant was design using the World Health Organization requirements.

Figure 11:1974 -SWDA

CONCLUSIONS

Panama Canal Water Filtration Plants have proven their effectiveness on providing drinking water through 100 years to different stake holders starting with the builders of the Panama Canal, The Panama residents, Military personnel, tourists and labor forces. This is done despite political and administrative changes.

Therefore Miraflores Filtration Plant methodology for treating water has proven to be effective during these periods. This has contributed in addition to other sanitary programs to maintain the country and the Canal Operations from waterborne diseases that may cause epidemics.

Waterborne diseases have been controlled from the beginning of the century due to the awareness of providing suitable drinking water to the population of Panama City and Colon and the workforce of the Panama Canal. This event has contributed silently the continued operation of the Canal and its service to the world without facing outbreaks that had occurred in the world.

The above is a great triumph taking in consideration that many of the epidemics or outbreaks have had their origin due to the maritime industry since they were transported by the ships.

Since the Plants were part of the Panama Canal Commission (Federal Agency of the U.S. government) they complied with those regulations.

Miraflores filtration Plant still complies with the 21st century regulations and produce drinking water for an increased population as well as Canal operations.

Mendoza has supported the development of the west bank communities where Canal employees live with their families.

The water quality has mitigated the outbreak of waterborne diseases.

Water treatment process has helped to shape the development of this country. First the construction of the

Panama Canal, then the operation and demand within the cities, and nowadays the importance during the Panama expansion program. These milestones have been well reported through time. However, without making headlines the water treatment Plants and processes have been silently contributing to the success of these events.

.

The state of the art design can be improved and challenge any other new design.

Water treatment plant basic facilities have not changed much. The process to achieve the quality of the water has been more refined.

Water treatment process still on the move and the latest state of the art technology is far from being discovered.

Mendoza has complemented the water supply to the communities west of the Canal, increasing the probability of having healthy population.

REFERENCES

- ANNUAL REPORT OF THE GOVERNOR OF THE PANAMA CANAL 1915. FISCAL YEAR ENDED JUNE 30 1915. Washington Government Printing Office 1915. 574 pages.
- Bunker George "Water Supplies of the Panama Canal" The importance of These Installations Is Not to Be Judge by the Size and Output, but the International Aspect of Their Services."
- Canadian Society For Civil Engineers "Centennial Conference 1987 Proceedings" Section I & II Volume 2, Montreal, Quebec, May 19-22 1987
- Canal Zone Census and Related information , 1954
- Canal Record Volume VIII August 26 1914 to August 18 1915. Panama Canal Balboa Heights Canal Zone 1915. The Panama Canal Printing Press, Mount Hope, Canal Zone 1915
- Canal Record Volume VII August 27 1913 to August 19 1914. Panama Canal Balboa Heights Canal Zone 1914.The Panama Canal Printing Press, Mount Hope, Canal Zone 1914.
- Contraloría General de la República de Panamá. Instituto Nacional de Estadistica y Censo. Proyecciones. Panamá 2010
- David McCullough " The Path Between the Seas" The creation of the Panama Canal 1870-1914. Simon adn Schuster . New York.
- Kirkpatrick R. Z. "Water supply of Panama Canal under operating conditions" Engineering News, pages 782-784 April 27 1916, vol. 75 No 17.
- Panamá en Cifras 1994-1998, Panamá, Nov 1999
- Qasim Syed, Motley Edward, Zhu Guang "Water Works Engineering, PLANNING, DESIGN & OPERATION", PHI LEARNING , New Delhi, 2011 , 900 p.
- Ramirez Ivan (from Internet) Historical Analysis of Cholera in Latin America. Dec 17 2006. Department of Geography, Michigan State University Epidemiology 805 Dr. Michael Dip.

- Wells George (Res. Engr. Municipal Engineering, Panama Canal Balboa Heights, Canal Zone , Panama) " Municipal Engineering and Domestic Water Supply in the Canal Zone" Preliminary Municipal Engineering, Article from the Panama Canal Transactions of the International Engineering Congress, 1915 San Francisco, Neal Pub. Co. 1916 .) p. 155-187.

- Wells George "The New Water Purification Plant At Miraflores Canal Zone. (P.C. 628.1 W46 pam) p 238- Reprinted Article September 10 1914.

ACKNOWLEDGE

Special Thanks to Rolando Cochez , Library Administrator for the Panama Canal Authority Roberto Chiari Library. His valuable contribution in the retrieval of historical data from the Panama Canal Archives contributed in a superb manner to the delivery of this paper.

The authors are grateful.

WAITANGI WHARF AND PORT UPGRADE – PROVIDING A CRITICAL LIFELINE AT THE EDGE OF NEW ZEALAND

by

Mark Foster1, Tom Shand[1], Manea Sweeney[1] and Matt Blacka[2]
1 Tonkin + Taylor Ltd, Auckland, New Zealand; email: mfoster@tonkintaylor.co.nz
2 Water Research Laboratory, Sydney, Australia

ABSTRACT

One of our most remote communities, the Chatham Islands, sits 800km off New Zealand's east coast and is home to 600 people. The islands are serviced by a port which provides a lifeline for the community through the provision of every-day goods and export earnings. The port is at the end of its structural life and significant upgrades are necessary. The Waitangi Wharf Upgrade, a project of some $58 million, involves reclamation, dredging and the construction of a large breakwater. The NZ Dept. of Internal Affairs requested that the Memorial Park Alliance comprising NZTA, HEB, Downer, Tonkin + Taylor and AECOM deliver upgrade works.

The project progressed from concept design, through consenting and detailed design within 12 months; an extremely tight timeframe for a project of this scale and complexity. Extensive community engagement was undertaken throughout the process including requirements at the port and surrounding areas, existing coastal processes and likely effects of the development and options for social and environmental improvements. This built trust and established relationships, very important factors for a project this size impacting on a very small community (i.e. the construction team increased the total island population by 6%).

This paper presents an overview of the project, and the unique challenges and lessons learned by working in such a remote environment, including:

- The relative importance of the project to its community reframed port activities by highlighting the critical importance of port infrastructure We consider that there is a huge, albeit complex, opportunity for the industry to more strongly connect communities to their ports for the benefit of the industry;
- How we must be sensitive to the communities in which we work, and strive to build strong relationships for everyone's benefit – particularly so with remote projects; and
- That physical model testing can add significant value in fine tuning a design, and providing confidence in the end solution.

Keywords: remote construction, port, breakwater, xbloc, collaboration.

1. INTRODUCTION

The Waitangi Wharf and Port Upgrade, a project of some NZ$58 million, involves reclamation, dredging and the construction of a large breakwater in one of New Zealand's, most remote communities. The extreme remoteness of the project, and its relative impact to the small community, highlighted some of the typical challenges and opportunities coastal practitioners face. This paper presents some of the key lessons learned and opportunities highlighted by the project.

2. THE CHATHAM ISLANDS AND WAITANGI WHARF

The Chatham Islands, located 800km east of Wellington, are an archipelago of 11 islands rich with seafood) aquatic animals and birdlife –.

The islands are New Zealand's easternmost community, with some 600 residents inhabiting Chatham and Pitt Islands. The remainder of the Chathams form one of New Zealand's most remote and valued conservation reserves.

The Chatham Islands are located within the Chatham Rise, a relatively shallow geological plateau, surrounded by much deeper trenches. The Chatham Rise is New Zealand's most productive fishing ground.

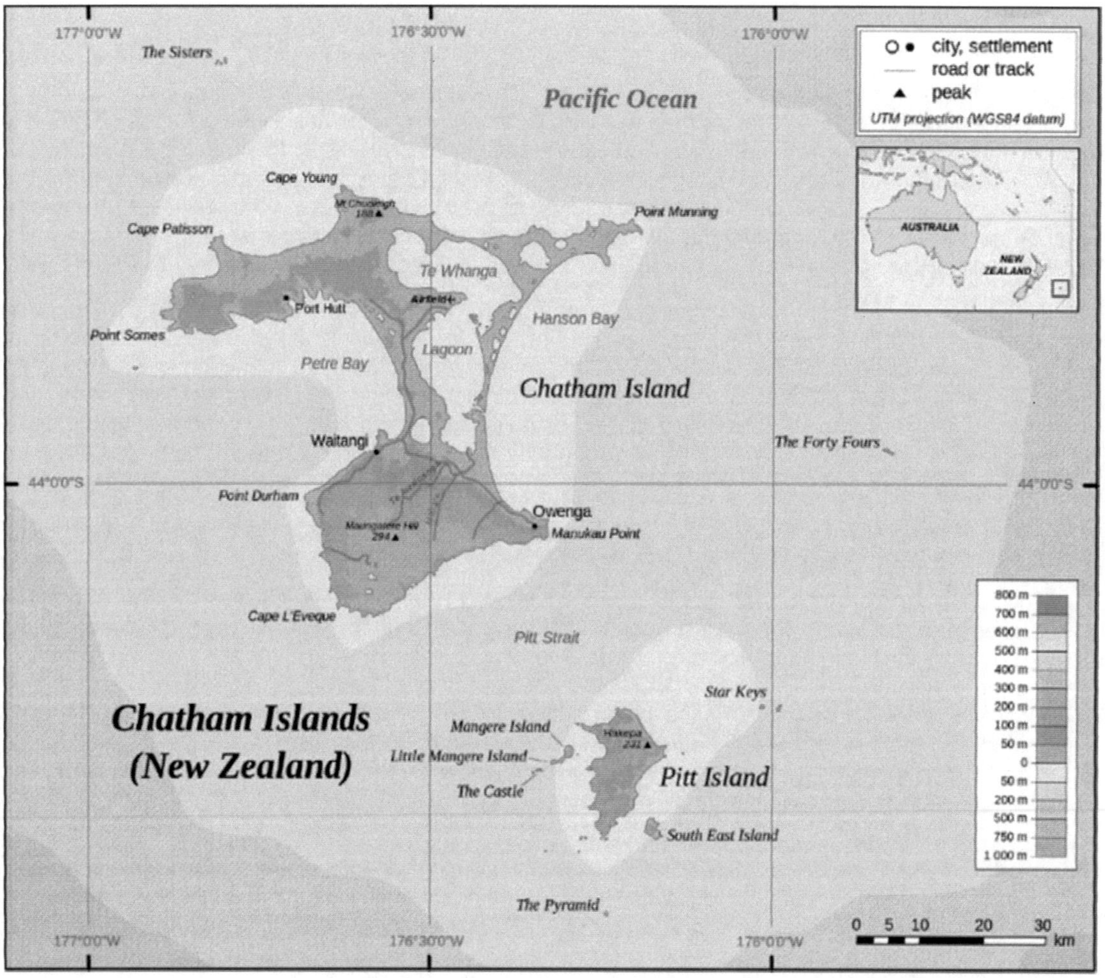

Figure 1 – Location of the Chatham Islands.

The Chatham Islands are economically and strategically important to New Zealand. They contribute around $46 million per annum to GDP, primarily through fishing, agriculture and tourism (Chatham Islands Council, 2015), as well as contributing to size of New Zealand's Exclusive Economic Zone (EEZ).

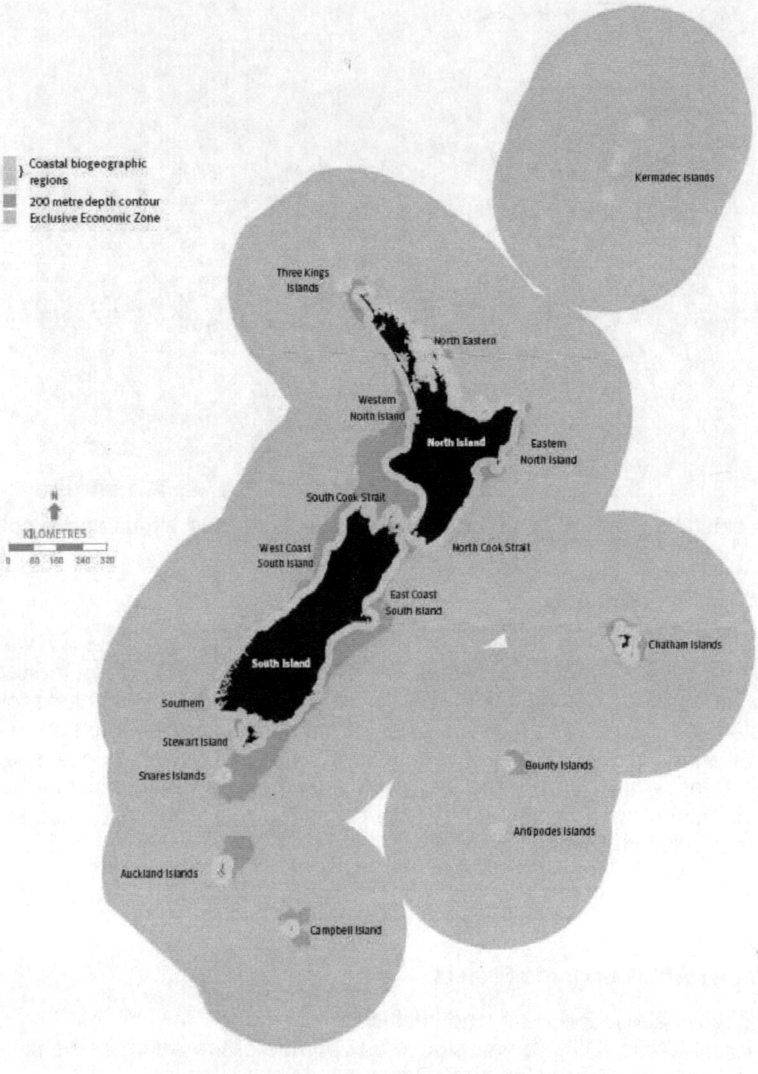

Figure 2 – New Zealand's Exclusive Economic Zone, illustrating the strategic importance of the Chatham Islands to NZ

The mainstay of the island's economy has always been centred on the natural resources of the sea, from historic practises such as whaling and sealing through to its present fishing activities that support the ongoing sustainability and economic viability of the island. The area within a 200 nautical mile radius of the Chatham Islands constitutes 10% of New Zealand's EEZ, with 6.3% of New Zealand's total fish catch being caught in this area.

The islands are serviced by one main port, located in Waitangi Bay. Waitangi Wharf is a critical lifeline – the only cargo-handling facility for exports and essential supplies, such as diesel for the electricity grid and fuel for air services. The existing wharf, and associated port is inefficient, and at times, unsafe.

The existing wharf and port layout consist of a T-shaped wharf leading from a restricted cargo handling area. Cargo handling operations are inefficient.

Of greater influence however, is the fact that the wharf is exposed to the large south-westerly swells from the South Pacific. Analysis undertaken as part of the project has shown that the wave motion at the existing wharf exceeds PIANC guidelines for safe operation approximately 50% of the time.

Figure 3 – The existing wharf at Waitangi showing the T-shaped wharf and small operational area.

2.1 Environmental conditions

The Chatham Islands are located at 44°S and exposed to consistent westerly winds and waves produced by the 'roaring 40s' weather systems. Waitangi Wharf is located in the southern corner of Petre Bay and in the lee of a rocky headland; Tikitiki Hill. This provides substantial protection from the predominant westerly swells as well as waves arising from strong northwest to northeast winds which are fetch-limited within the bay. Extreme wave height (100 year ARI; swell or local sea) were hindcast (Metocean, 2015) at less than 2.2 m, however period is long (>14s) and persistent background swell occurs year round. Tidal range is small (<1m) and, below the rock shelf that lines the coastline, the sandy seabed is relatively flat.

3. The Waitangi Wharf Upgrade Project

With repairs required, safety concerns and inefficient operations for the existing wharf, The NZ Department of Internal Affairs (DIA) commissioned an upgrade of the wharf, on behalf of the Chatham Island Council and Chatham Islands Enterprise Trust.

DIA requested that the Memorial Park Alliance – an existing partnership comprising The New Zealand Transport Agency, construction firms HEB and Downer, and consultants, Tonkin + Taylor and AECOM – deliver the $58 million worth of urgent upgrade works.

The resulting Waitangi Wharf Upgrade Project seeks to improve the safety, reliability and efficiency of wharf operations. And to future-proof the wharf for the community's expanding export and production requirements, while improving animal welfare for exported livestock.

The key elements of the project include:

- Reclamation of 9,500m2 of land for a new por;t
- Construction of new commercial and fishing wharves, 90m and 40m respectively;
- A 180m long breakwater;
- Dredging of the approach and berthing area;
- Establishment of two quarries to provide aggregate for reclamation fill and breakwater armour; and
- Establishment of a concrete batching plant and construction yard to produce the three thousand concrete armour units required to withstand the local wave climate.

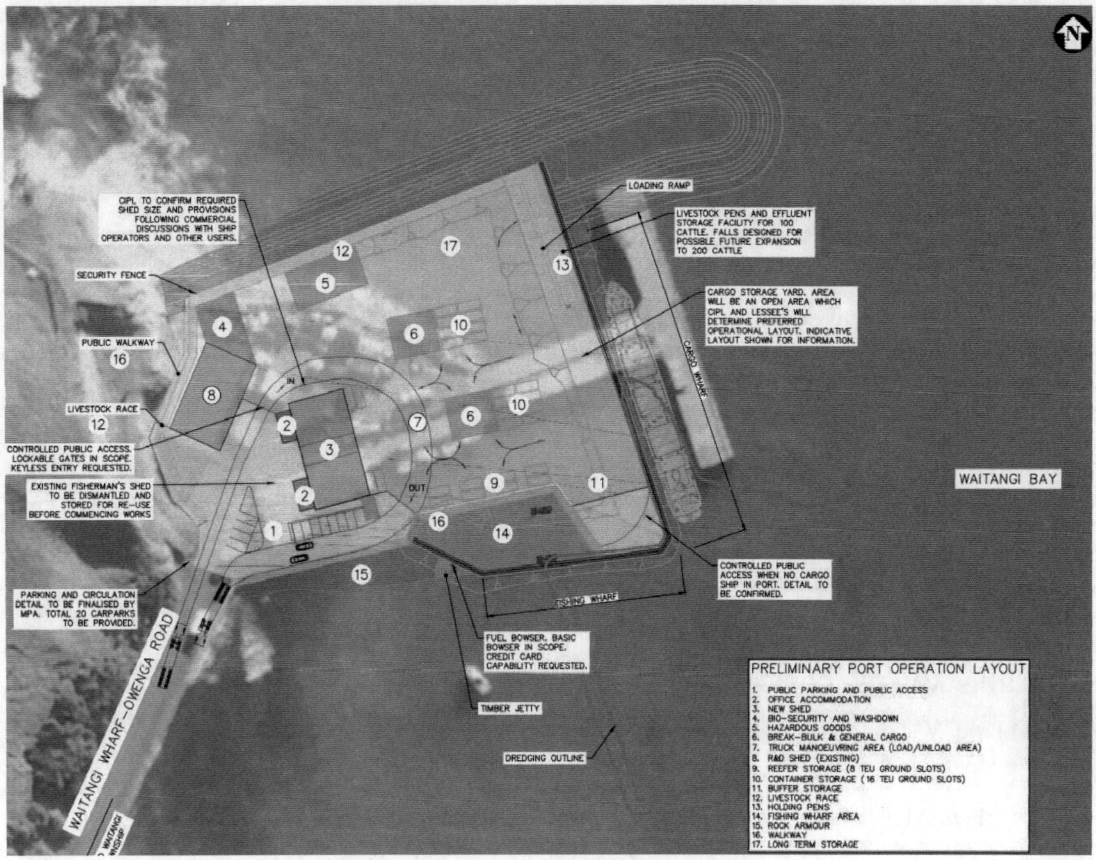

Figure 4 – Layout of the proposed Waitangi Wharf. Showing the key features relative to the existing situation. Note to reviewer – better quality image with readable text will be sourced.

4. THE PROJECTS IMPORTANT CONTEXT

Some of the lessons learned came about as a result of the unique Chatham Islands context which is important to understand. That particular context was;

4.1 A small and tight knit community

The island population is just 600 people. And they are very parochial, with most resident's generational islanders.

The project was going to result in a temporary workforce meaning a 6% increase in the islands population. We had to be very conscious of the potential impact this would have on the small, tight knit community.

4.2 A community where everyone has a direct relationship with the port

Individual's reliance on the port, and relationship with it, is so much more direct on the Chathams than in mainland New Zealand.

Almost every family on the island is involved in farming or fishing, often both. Their livelihood relies on the individual farmer or fishermen arranging their own shipping off the island direct with the shipping companies.

As consumers, the relationship with the port is so much more direct. If a Chatham Islander needs to purchase a new fridge – they cannot go to the shops. They must buy one from 'New Zealand' (as locals refer to the mainland) and arrange for it to be shipped to the island, where they will pick it up from the port themselves.

Such is their reliance on the port (and the unreliability of shipping) that locals talk about the shipping schedule like most farming communities talk about the weather.

4.3 A remote location with limited infrastructure

Constructing such a large project, on an island with limited infrastructure presents many challenges. Such as;

- No suitable quarry for the 100,000 cu.m of aggregates required;
- No rock large enough for breakwater armour;
- No concrete plant to construct armour units;
- No water supply to make concrete (the town supply was already under capacity);
- Very limited, and only small, construction plant on the island;
- A diesel supply chain set up to support the local community, and not the large amount of heavy machinery that would be needed to construct the works;
- Limited accommodation for the 40 strong workforce;

We were required to effectively engineer the project from first principals. This necessitated close collaboration between the client, design and construction teams, the local community and local businesses.

5. BREAKWATER DESIGN

As part of the Waitangi Wharf upgrade, a breakwater was deemed necessary to:

- Provide protection to the reclamation against wave erosion;
- Provide protection to the reclamation against wave overtopping and inundation;
- Reduce the wave climate in the lee of the breakwater to decrease the number of days where berthing is not possible at the wharf.

5.1 Breakwater planform

Ship operators suggest that with the typical 12-14s period swells, they are currently able to berth in swell wave conditions less than 0.5 m, berthing becomes marginal in waves of 0.5 - 0.75m and they are not able to berth in wave heights larger than 0.75m. However, mooring lines are reported frequently broken indicating berthing safe conditions are being frequently exceeded.

The new breakwater and wharf configuration will, however, induce waves to approach the moored vessels from the quarter (45°) to beam (90°) rather than head on (0°) as presently occurs. Vessels are less stable with waves approaching from this angle [5] and so require a lower wave climate to achieve the same level of service. Additionally, the proposed berth is a vertical quay wall which will reflect incident waves further reducing the tolerable wave condition under which berthing and working is safe compared to the existing open piled structure. These changes must be taken into account along with the reduced wave height in determining any change in berthing under a new wharf configuration.

A combination of phase-resolving wave modelling (CGwave and Mike21 BW) and Optimoor was used to determine ship motion under the existing configuration and for incrementally increasing breakwater lengths. PIANC (1995) provides recommended motion criteria for safe working conditions and gives the effective thresholds for the ship motions of Surge, Sway, Heave, Yaw, Pitch and Roll for various vessel classes. Analysis found that under the existing configuration, ship motion was exceeding safe working conditions 66% of the time or an average of 246 days/year (c.f 11% of the year based on the operator's Hs=0.75m threshold). Increased breakwater length resulted in reduced wave height at the berth, a length of 65 m beyond the wharf was eventually adopted based on cost and operability criteria. This length reduced ship motion exceedances to 26% of the time based on PIANC (1995).

5.2 Breakwater armour

Breakwater armour protects the breakwater structure from wave attack and was a critical component of the project. Due to the very remote location, a 'zero maintenance' (or as close to as possible) was sought.

Early inspection and testing of local rock by engineering geologists indicated that the potential rock quarry sources on the island were good quality basalt with densities in excess of 2.8 t/m3, but size was limited by shear planes to typically less than 400kg. Design was therefore undertaken with this constraint in mind.

A range of options were considered including conventional rock breakwater, reshaping breakwater, concrete caisson structure or concrete armour units. A lack of suitable size rock on island and very high transport costs to island resulted in concrete armour units being preferred for breakwater construction. After reviewing a range of potential armour units as described within PIANC (2005), Xbloc® units have been selected as most cost-effective and well-proven unit.

Due to the requirement for 'low maintenance', a 0.1% AEP wave height was selected incorporating the upper 95% confidence interval [2]. This resulted in a design wave height of Hs=2.7 m (the site is in the lee of a large headland). The resulting Xbloc® (DMC, 2014) was 0.75m3 or 1.8T and small enough that the available rock on island was suitable to be used as underlayer material (a critical cost factor), something not possible with larger randomly placed units.

Figure 5 - Xbloc placed on a test slope

5.3 Physical model testing

Physical model testing of armour stability was undertaken at the Water Research Laboratory in Sydney on a 2D breakwater cross section and around the breakwater head using a quasi-3D test. The testing also allowed wave forces on the crown wall and wave overtopping rates to be assessed. Results [6] were used to optimise design and identify value engineering opportunities such as replacing crest armour for local rock and also to identify potential design issues.

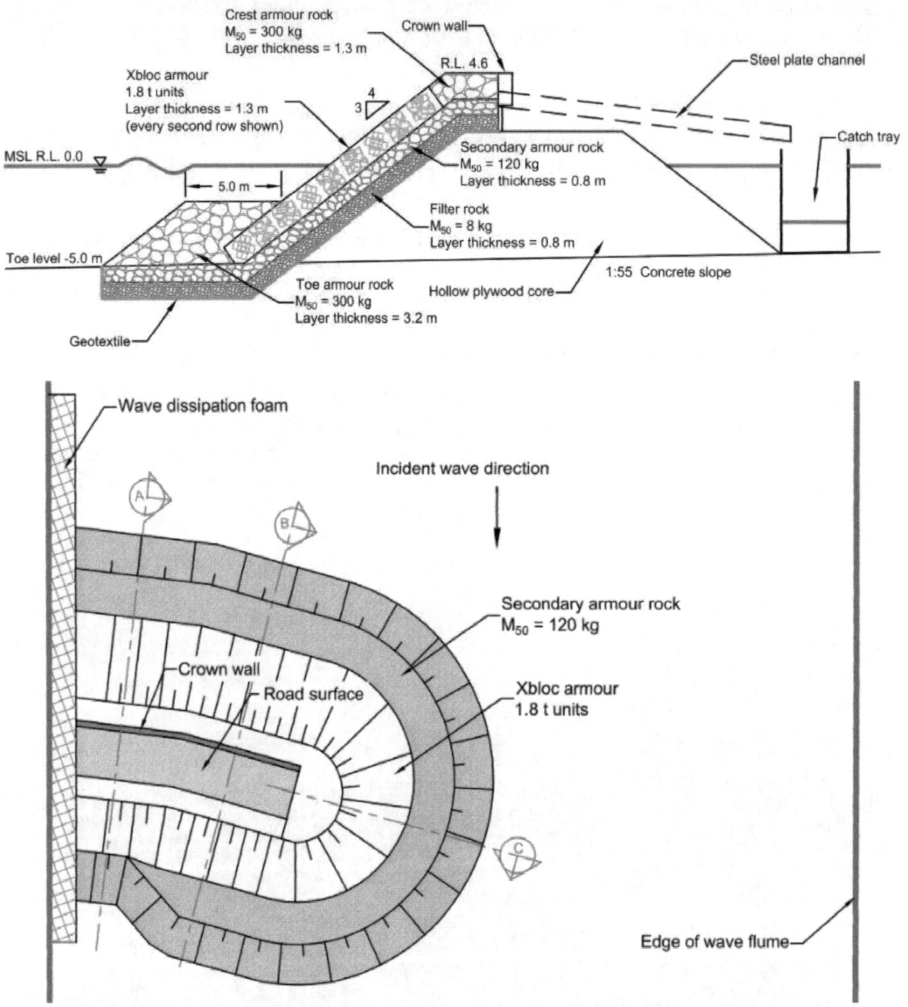

Figure 6 - Example of physical model 2D (top) and quasi-3D (lower) test configurations

Some such issues included a lack of crown wall return at the end of the breakwater allowing Xbloc® to be displaced across the breakwater and displacement of toe armour on the leeward side of the breakwater head (a known weak area). Toe armour rock were displaced beyond expectation based on empirical guidance resulting in Xbloc® being exposed and displaced from the toe and beginning to unravel (a 'catastrophic failure'). These problems were solved by modifying the design; the crown wall was continued around the end of the crest and the Xbloc® toe was continued to underlying rock rather than rafted on sand. This testing provided confidence that the design was both robust and as cost-efficient as possible.

5.4 Construction aspects

As previously discussed, the typical wave climate was low but persistent year round (refer Figure 7). This meant that construction would need to occur during wave events. This was challenging given the small core and filter materials and precise placement required.

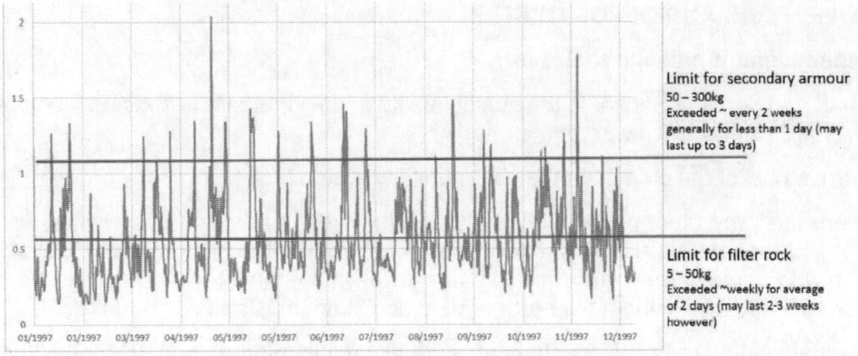

Figure 7 – Time-series of hindcast wave height [5] for 2012 with indicative thresholds for initiation of damage to core/filter layer and secondary armour.

To improve certainly around the design, physical model testing was undertaken on 'in construction' sections to simulate typical conditions and moderate storm events. It was found that the core and filter material were moved under waves as low as 0.5 m, which occurred weekly for 2-3 days but at times for as long as 2-3 weeks. The secondary armour was moved when waves exceeded 1.2m which occurred every 2 weeks on average, though generally for less than one day. In both these cases, the damage was typically a slumping of material (Figure 8) from above the waterline to below the zone of wave energy. This provided useful information on the construction thresholds and on repair methodologies.

Figure 8 - Effect of increasing wave height from 0.55 to 1.2 m on secondary armour rock before the xbloc are placed

6. KEY LESSONS AND OPPORTUNITIES

6.1 Connect communities to their ports

Waitangi Wharf and port is essential to the sustainability of the Chatham Islands as a community. And the community are intrinsically aware of this.

The short-term impact of the project on the community was significant.

Yet much more than any other project we have been involved in before, there was an understanding that those impacts, and potential impacts, were part of the trade off. The community were right to have concerns, and we worked closely with them to minimise any adverse effects. But they knew the project was for the good – because without the port, there is no Chatham Islands community.

It could be argued that any city in New Zealand is equally reliant on their port. Certainly the country as a whole is equally reliant on its ports in totality. We must be able to import and export. Yet, on mainland New Zealand, the vast majority of the public do not appreciate this importance. And as such, ports most often face resistance to development and growth.

If we can connect our mainland communities to the importance of our ports, we could see a shift in sentiment towards port development and port operation. The authors believe there is a huge, albeit complex, opportunity for the industry in this regard.

6.2 Procurement matters

Too often remote projects, particularly in the Pacific, are procured using a linear procurement model – of design, tender and construct.

The New Zealand Transport Agency (NZTA), as advisors to DIA, are a mature infrastructure delivery agency. They recognised that with the tight timeframes, limited scope definition, and the challenges of delivering a project in a location with limited infrastructure, that an Alliance procurement model was the best approach.

Alliancing brings all parties required to deliver a project, including the client, into a single collective agreement. With a pain share / gain share agreement. The model encourages all parties to work together for best for project outcomes.

From the outset the environmental, design and construction teams were working side by side. Even the earliest decisions were made with all delivery partners present. For a project where all of the plant, machinery and person power required to undertake the works needed to be shipped more than 800km – such collaboration was priceless. Design decisions were influenced by the plant that was going to be present on the island.

Its known that good collaboration often leads to better outcomes. In the case of remote projects, this is even more acute. Funders should consider procurement methods carefully for remote projects – as there is significant advantage in choosing the right method, and one that allows collaboration, as DIA and NZTA did in this instance.

6.3 Good relationships help to achieve a win-win

The Waitangi Wharf Upgrade really put the value of relationships under the spot light.

With such a small community, who were going to be significantly impacted by the works, and whom we needed considerable assistance from, we recognised early that relationships would be paramount.

The Alliance dedicated considerable effort to integrating with the community and building strong relationships.

One of our drivers was that the project relied heavily on the services of the islands businesses. Be it, accommodation, fuel, shipping, flights, and labour force – we needed them to deliver the project. We needed them more than then needed us.

But with a small community, and given the economic impact the project was likely to have, we had to be sensitive around how we engaged with the business community.

Ultimately, our philosophy was to strive for win-win outcomes, with as many local businesses as possible.

Another driver was our will to have a positive influence on the island community. With such a small and tight knit community, the Alliance workforce had the ability to have a significant impact on the island. We needed to make sure it was a positive, and respectful impact. Considerable effort was put into inducting all staff into the culture of the island – with our induction packs put together with the help of the local community. And we recognised that the everyday counted. It was important for the workforce to be part of the community, from entering a touch team in the local competition, to attending community events, and even frequenting the local pub – the was a very conscious effort to be a 'good island citizen'. It was one of the non-financial key result areas for the project.

To date, our efforts in building relationships have been extremely worthwhile. The Alliance has achieved better financial outcomes through agreements with the business community. The community feel they have a project team they can trust. The workforce are enjoying their time on the island. And the island are happy to have their temporary residents contributing to the island community in a positive way.

This would not be possible with investing the time to really get to know the local community and build those strong relationships.

Figure 9 - Waitangi Wharf upgrade during construction in Jan 2017

7. SUMMARY

The unique challenges and context of constructing the $58M Waitangi Wharf Upgrade on the Chatham Islands, has served to highlight opportunities that coastal practitioners and port developers may be able to seize with other developments. Most importantly;

- The relative importance of the project to its community reframed port activities by highlighting the critical importance of port infrastructure We consider that there is a huge, albeit complex, opportunity for the industry to more strongly connect communities to their ports for the benefit of the industry.
- The challenges of remote working are best overcome with a highly collaborative approach. And procurement plays a key role in fostering collaboration.;
- How we must be sensitive to the communities in which we work, and strive to build strong relationships for everyone's benefit – particularly so with remote projects; and
- That physical model testing can add significant value in fine tuning a design, and providing confidence in the end solution.

8. REFERENCES

[1] DMC (2014) Guidelines for Xbloc® Concept Designs. Delta Marine Consultants, March 2014.

[2] Metocean (2015) Metocean Summary: Waitangi Wharf, Chatham Islands. Report prepared for the Memorial Park Alliance, January 2015.

[3] PIANC (1995) Criteria for movements of moored ships in harbours - a practical guide. MarCom Working Group 24.

[4] PIANC (2005) Catalogue of Prefabricated Elements. MarCom Working Group 36.

[5] Thoresen, C.A. (2003) Port Designer's Handbook: Recommendations and guidelines. Institution of Civil Engineers, January 2003.

[6] Water Research Laboratory (2016) Chatham Islands Port Waitangi Physical Modelling. WRL Technical Report 2015/17 by D. Howe, I. Coghlan, J. Carley and M. Blacka.

TOWARDS A FRAMEWORK FOR INTEGRATED, ECOSYSTEM-BASED PORT DEVELOPMENT

by

P. Taneja[1], T. Vellinga[1], K. Appeaning Addo[2], J. Slinger[3], H. Vreugdenhil[1,3], W.P. de Boer[1,3], A. Kangeri[4], B. Kothuis[1], M. Koetse[5], L. Hagedoorn[5], C. van Dorsser[1], E. Mahu[2], B. Amisigo[2]

ABSTRACT

This paper describes an integrated, ecosystem-based research project exploring what it means to be a sustainable port in a developing country context. Africa is one of the few areas in the world where greenfield port development is still occurring in addition to brownfield development. This means there are many challenges and opportunities to design for sustainability and to nudge existing ports towards more sustainable activities. A stepwise, case study oriented approach to tackling these issues is explicated in an effort to understand the advantages for port developers and their financiers to move in this direction.

1. INTRODUCTION

A paradigm shift is required in our approach to large scale infrastructure development including ports, whereby the emphasis lies on achieving our objectives in an ecosystem and societal context in an uncertain environment. There is a growing recognition of the need for more sustainable approaches to port development aimed at balancing social, environmental and economic aspects. In spite of this, an integrated inter-disciplinary approach to sustainable port development, which embraces the four perspectives of engineering, ecosystem services (ecology and economy) and governance is lacking. The ongoing NWO-WOTRO UDW project "Integrated and Sustainable Port Development in Ghana within an African context" (NWO, 2018) addresses this gap.

The project which started in May 2016 spans 3 years and is carried out by a multidisciplinary consortium, wherein academia, applied research institutes, knowledge institutes, practitioners and potential users collaborate as partners and interact with a broad range of local stakeholders. It adopts an inter-disciplinary approach that integrates different aspects of sustainable port development to create a knowledge base serving a design-framework and accompanying design tools that can be applied in developing more sustainable ports. Using a case study in Ghana as the central application focus, subsequent replication of new knowledge and practices is expected within Ghana and beyond, followed by eventual institutionalization through bottom-up adoption and activated Pan-African networks.

2. TOWARDS SUSTAINABLE PORTS IN AFRICA

The selected case study is the Port of Tema, located in the south eastern part of Ghana along the Gulf of Guinea. Tema has evolved from a small fishing village in the mid twentieth century to become Ghana's leading industrial centre and seaport in the last decade. The port is undergoing expansion to serve rising cargo traffic volumes as Ghana's economy maintains its high rate of growth. To integrate knowledge and learning on the case study, four sub-projects focussing on engineering (P1), ecology (P2), economy (P3) and governance (P4) have been defined.

[1] Delft University of Technology, Faculty of Civil Engineering & Geosciences, & Faculty of Technology, Policy & Management, Netherlands, j.h.slinger@tudelft.nl
[2] University of Ghana, Department of Marine and Fisheries Sciences, Ghana
[3] Deltares, Netherlands
[4] Wageningen Marine Research, Netherlands
[5] Institute for Environmental Studies, Netherlands

P1 *Port engineering and design* views the port as a component within the coastal zone and develops alternative layouts for the port expansion of Tema. Hereby, the principle of Building/Working with Nature is gainfully applied in exploring infrastructural options to maintain, restore or offer opportunities to coastal ecosystems. The changes engendered by alternative port layouts and diverse port infrastructural designs are studied in terms of their ecological and societal effects.

P2 *Marine ecosystems and coastal erosion* examines the dynamic interactions between infrastructural elements and the coastal ecosystem. Here the emphasis lies not on the infrastructural elements, but on the ecological components, their functioning and evolution from the start of port development in Tema. Understanding of the changes in the functioning of the ecosystems are sought so that opportunities to create added value for nature from the the port can be identified. For instance, by countering erosion problems and integrating restoration of ecological habitats in new road infrastructure development.

P3 *Economic valuation of ecosystem services* applies state-of-the-art methods to quantify the evaluation of changes in ecosystem service delivery to humans. This allows the inclusion of ecosystem services in the socio-economic analysis of port development alternatives. Social aspects identified as relevant in P4 will also be valued.

P4 *Governance of green port development* formulates an opportunity-oriented implementation model for port development in which port authorities, contractors, nature conservation organizations, ecologists, engineers and local stakeholders cooperate to formulate a better fit between ports and their social-ecological landscape. In this process, the values that stakeholders hold regarding the future alternative states of the port-city and its surroundings are accessed.

The overall project encompasses the following steps:

- gathering relevant data for the pilot project through desk studies, field research and workshops;
- exploratory studies and brainstorm sessions to formulate alternative port layouts and identify potential ecosystem services;
- stakeholder-inclusive workshops to elicit stakeholder values and local knowledge;
- integration of the findings into alternative conceptual port layouts. These alternatives include: status quo as reference layout, the current port expansion plan of Ghana Ports and Harbours Authority (GPHA), an incremental port development plan with a focus on value addition by considering sustainability, and an innovative layout;
- detailed assessment of alternative port layouts with respect to functionality of port operations, morphological effects, environmental impacts on the marine and coastal ecosystems, and economic valuation of ecosystem services;
- a detailed assessment of stakeholder values in relation to port-city development and the effects on surrounding ecosystems;
- the formulation of a framework for sustainable port development, tested in Tema, Ghana.

The pilot case study will result in a port design for Tema that complies with the requirements as to functionality and sustainability. The knowledge and insights gained from the case study are also used to develop a potentially generic framework and tools for stakeholder-inclusive design of sustainable ports. Knowledge dissemination in accordance with a well-formulated research uptake strategy forms an important component of the project, as does the desire to test the developed framework by applying it on other ports in Africa and elsewhere.

3. CONCLUDING

By describing this on-going research project, including the underlying vision, the challenges and the objectives, the activities pertaining to the selected pilot project in Ghana, as well as the envisaged end results, this paper aims to share new knowledge and learning about the practice of port development. Finally, as the approach and underlying principles are applicable to large scale infrastructure development at other locations, we encourage others to apply these principles in the sustainable development of ports.

References

NWO (2018). Integrated and Sustainable Port Development in Ghana within an African context, Urbanising Deltas of the World Program, URL: http://www.nwo.nl/en/research-and-results/research-projects/i/95/13995.html ; accessed February, 2018.

Acknowledgements

The authors wish to express deep appreciation to the Netherlands Organisation for Scientific Research (NWO-WOTRO) for financing this research project (nr. W07.69.206) under the Urbanizing Deltas of the World programme.

DESIGNING FOR STAKEHOLDER VALUES IN PORT DEVELOPMENT IN AFRICA

by

J. Slinger[1], P. Taneja[1], B. Kothuis[1], H. Vreugdenhil[1,2], W.P. de Boer[1,2], A. Kangeri[3], M. Koetse[4], L. Hagedoorn[4], C. van Dorsser[1], E. Mahu[5], B. Amisigo[5], K. Appeaning Addo[5], T. Vellinga[1]

ABSTRACT

This paper addresses the need for stakeholder-inclusive design in sustainable port development. This involves learning about the values and knowledge resources of local stakeholders at an early stage, but is a step that is often omitted in current port planning processes. However, it is essential in creating added value and in avoiding costly delays when port development is stopped or delayed by social impact findings at a later stage. A stakeholder engagement process for the Port of Tema, Ghana, is used to illustrate the types of activities required and offer insights in the outcomes. In particular, a game structuring method applied in a 50-people workshop in February 2017 demonstrated that the expertise of local stakeholders and insight into their preferences regarding potential futures for the port city and its surroundings can inform planners, port authorities and engineering scientists about what it means to be a sustainable port in a developing country context.

1. INTRODUCTION

Sustainable development recognizes that growth must be both socially inclusive and environmentally sound to meet the needs of present and future generations. While seaports provide an essential link to the world market, enabling countries to trade their goods and strengthen their economy, these global and regional benefits do not always translate into benefits to local stakeholders. Indeed, there is often a mismatch between the positive global and regional effects of port development and the impacts (dis-benefits) that can be experienced locally (Rodrigue, Comtois and Slack, 2017; Merk, 2013). Port development requires a paradigm shift from a traditional approach to a stakeholder-inclusive, integrated approach to be sustainable (Vellinga et al., 2017; Schipper et al., 2017).

The Ghanaian Port of Tema forms the focus point for an ongoing project "Integrated and Sustainable Port Development in Ghana in an African Context" (NWO 2018). The research project is carried out by a consortium comprising the University of Ghana, and Dutch knowledge institutes and private sector companies. While the broad research aims are to gain insights in how to develop integrated and sustainable African ports that meet economic, social and environmental demands, one of the specific objectives is to find out what local stakeholders value as a sustainable port, in that it (i) creates value for stakeholders working and living in the immediate environment of the port, (ii) can still meet demands in 50 years, and (iii) can grow apace with economic developments of the future. The purpose of this paper is to provide tangible examples of the range of stakeholder-inclusive, value-based interactions implemented in the port city of Tema.

2. THE STAKEHOLDER ENGAGEMENT PROCESS

Recognising that stakeholder-engagement is a prerequisite for sustainability, four activities were initiated. The first activity involved organising a stakeholder workshop prior to the inception of the project. This occurred in Tema in 2015, and the outcomes were used to refine the focus of the research project and ensure that it addressed issues raised by the workshop participants. The

[1] Delft University of Technology, Faculty of Civil Engineering & Geosciences, & Faculty of Technology, Policy & Management, Netherlands, j.h.slinger@tudelft.nl
[2] Deltares, Netherlands
[3] Wageningen Marine Research, Netherlands
[4] Institute for Environmental Studies, Netherlands
[5] University of Ghana, Department of Marine and Fisheries Sciences, Ghana

participants included local, regional and national authorities, industry, port developers, tourism and nature interest groups, amongst others. Participants first played the serious game Port of the Future. This is an interactive game in which players assume the roles of various stakeholders and decide on measures to be implemented during the process of port development. Each round in the game involves a negotiation within and between the different roles. The envisaged learning relates to the potential sustainability of measures and the importance of negotiating agreements on measures. The game served as a communication tool to bring across the significance and implications of sustainable port development for stakeholders, and helped to deepen understanding of different and often conflicting interests while attempting to create a port that not only serves economic functions but is in harmony with the surrounding ecosystems. Following the game, participants were encouraged to elaborate and discuss the issues they considered of relevance for Tema. The outcomes were an increased awareness of sustainable port development, and concurrence amongst the participants on the economic, ecological and social aspects that the project should address. These outcomes were included in the "Integrated and Sustainable Port Development in Ghana in an African Context" project proposal (NWO 2018).

Next, a workshop for project partners and supporting parties (see Vellinga (2017) for information about the project organization) was held in Delft, the Netherlands, in September 2016 with similar objectives (Vreugdenhil, 2016). The Ghanaian researchers participated in the workshop through an active skype link. The workshop agenda incorporated an introductory presentation on the project, which included the input from the first workshop, a question and answer session, a personal mini-survey over what is (or is not) a sustainable port in Tema or an ideal world. This was then followed by group sessions on sustainable ports and general discussion. Then, the serious game Port of the Future was played as part of the workshop. This was done for parity with the stakeholder workshop in Ghana. Finally participants were asked to reflect on workshop. The outcomes were similar to the first workshop in Tema, namely increased awareness of sustainability and consensus on the most important issues to address via the project. In addition, the connectedness of the network around the project was strengthened, and the inputs and insights from the Ghanaian partners added depth to the discussion on sustainability within an African context.

Following an intensive six-month period of data accumulation and research, a multi-stakeholder workshop was organized at the Ghana Ports and Harbours Authority Headquarters (GPHA Tower) in Tema in February 2017. This crucial activity in the overarching research project was designed to engage local port-related stakeholders in an innovative, value-based exploration of the future of Tema and its port. Clearly, the selection of participants and the method employed to engage with stakeholders are critical to success here. Invitations to stakeholders were issued on the basis of their involvement with the port. So, participants were drawn from different layers within the GPHA, and ranged from freight forwarders, local businesses involved with the port, port developers, to interest group representatives for tourism and the coastal ecosystem, as well as local scientists. More than 50 people attended the workshop, participating thoughtfully and enthusiastically in the different steps in the process. A six step game-structuring methodology (see Box 1) underpinned the workshop (Cunningham et al. 2014; Slinger et al. 2014).

The multi-stakeholder workshop was grounded in a wider stakeholder-inclusive situation assessment. Because the multi-stakeholder workshop was oriented to port-related stakeholders, an additional process of interviewing local residents with no direct relationship to the port was initiated in February 2017. This process was continued in 2018, resulting in more than 25 interviews shedding light on the effects of port development on those who live in, or near, Tema. These include traditional leaders, the original inhabitants of Tema prior to Port construction, local fishermen, teachers, immigrants to the area, long-time residents, town-planners and many more. The effects on these stakeholders are both positive and negative and vary in their impact on the lives of the people concerned. The envisaged output of this process is a deep understanding of the long term consequences of port expansion in all its facets, captured in a book that is currently under preparation.

Game structuring is a transdisciplinary practice developed specifically for environments in which multiple stakeholders, holding diverse values, are involved in the long term management of a social-environmental system, such as Tema and its Port. The purpose is thus not to reach consensus nor to solve conflicts, but rather to explore the existing different interests, values, potential and preferred futures, as well as the potential clusters of actions belonging to these futures and the associated restrictions, objections and hurdles. The workshop method rests on operations research traditions, using game theory in combination with techniques such as systems modeling and community narrative generation in particular. This potentially offers the participants in such a long term social-environmental negotiation a wider view on their own and others' positions, both now and in the future. The aspiration is that this leads to commitments to joint action based on shared values on the part of participants.

Workshops based on the game structuring approach have been conducted in South Africa (Slinger et al., 2014), the Houston Galveston Bay Area (Cunningham et al. 2015), and the island of Texel in the Netherlands. In all these cases, the method was experienced positively by participants and has been associated with increased stakeholder engagement in local decision making processes in the subsequent months

The game structuring, value-based design method as applied in Tema, Ghana entailed six steps:

1. Getting acquainted
2. Developing the system story; local stakeholders on past, present and future of Tema and its Port
3. Developing the system story; researchers on Sustainable Ports in Africa, Tema and its Port
4. Identifying key stakeholders
5. Developing visions
6. Voting on visions from the point of view of key stakeholders.

Documentation on the workshop process and outcomes was subsequently supplied to all participants.

Box 1: Game structuring method for stakeholder engagement workshops

3. OUTCOMES OF STAKEHOLDER ENGAGEMENT

The four activities comprising the stakeholder engagement process of the "Integrated and Sustainable Port Development in Ghana in an African Context" project to date have yielded:

- a composite story of the evolution of Tema from before the first port was constructed, to the present day and beyond;
- positive and negative visions for the future of the port and city of Tema;
- insights on the values held by local stakeholders in relation to these visions of the future;
- understanding of effects of the port development on the lives of local stakeholders.

4. CONCLUDING

The composite understanding of stakeholder values and port effects has enabled the subsequent identification of ecological, economic, infrastructural and social components that could potentially be used in the sustainable development of Tema and its port. In summary, grounded stakeholder engagement methods have proved vital in accessing local knowledge on how to improve the fit between the port of Tema and its social-ecological environment. We anticipate that local stakeholder inclusion and value-based design will prove equally important, and form a generic process-element, in a framework for sustainable development of ports in other localities in Ghana, Africa, and worldwide.

References

Cunningham, S. W., Hermans, L. M., Slinger, J. H. (2014). A review and participatory extension of game structuring methods. EURO Journal of Decision Processes, (DOI) 10.1007/s40070-014-0035-8

Cunningham, S.C., Kothuis, B.L.M., Slinger, J.H. (2015). Contested Issues Game structuring Approach (CIGAS). Game theory, real actors and values in the Houston Galveston Bay region. In: Kothuis et al. (eds.) *Delft Delta Design. Houston Galveston Bay Region. Texas, USA*. Pp. 58-59

NWO (2018). Integrated and Sustainable Port Development in Ghana within an African context, Urbanising Deltas of the World Program, URL: http://www.nwo.nl/en/research-and-results/research-projects/i/95/13995.html ; accessed February, 2018

Merk, O. (2013). The competitiveness of global port-cities: synthesis report. *OECD Regional Development Working Papers*, 2013(13)

Rodrigue, J. P., Comtois, C., & Slack, B. (2017). *The geography of transport systems*. New York, Routledge. ISBN 978-1138669574

Schipper, C. A., Vreugdenhil, H., de Jong, M. P. C. (2017). A sustainability assessment of ports and port-city plans: Comparing ambitions with achievements. *Transportation Research D* 57: 84-111. http://dx.doi.org/10.1016/j.trd.2017.08.017

Slinger, J. H., Cunningham, S. W., Hermans, L. M., Linnane, S. M., Palmer, C. G. (2014). A game-structuring approach applied to estuary management in South Africa. EURO Journal of Decision Processes. (DOI) 10.1007/s40070-014-0036-7

Slinger, J.H, Kothuis, B., Vreugdenhil, H., van der Kraan, A. (2017). Stakeholder Workshop, Tema, Ghana. Sustainable Port Development in Ghana within an African Context, NWO UDW Project W 07.69.206, 9 February 2017.

Vellinga, T., Slinger, J.H., Taneja, P., Vreugdenhil, H. (2017), Integrated and Sustainable Port Development in Ghana, SMTC 2017, 26-28 April, 2017, Singapore.

Vreugdenhil, H., Slinger, J.H., Keijer, H. Kothuis, B., Alink, A. Taneja, P., Vellinga T. (2016). Kick off Meeting and First General Workshop for Supporting Parties, Sustainable Port Development in Ghana within an African Context, NWO UDW Project W 07.69.206, 20 September 2016.

Acknowledgements

The Netherlands Organisation for Scientific Research (NWO-WOTRO) is acknowledged for financing project nr. W07.69.206 under the Urbanizing Deltas of the World programme. Our sincere gratitude is extended to all participants in the stakeholder engagement processes related to the Port of Tema.

THE CONTRIBUTION OF NATURE-BASED CONCEPTS TO SUSTAINABLE PORT DEVELOPMENT

by

J. Slinger[1], D. Rijks[2], T. Vellinga[1], P. Taneja[1], H. Vreugdenhil[1,3], W.P. de Boer[1,3], A. Kangeri[4], B. Kothuis[1], M. Koetse[5], L. Hagedoorn[5], C. van Dorsser[1], E. Mahu[6], B. Amisigo[6], K. Appeaning Addo[6]

ABSTRACT

The potential added value of ecosystem-based concepts in port design are explored using a series of examples from the Netherlands and Ghana. The first example relates to the sandy dunes comprising part of the protection of the Maasvlakte II extension to the Port of Rotterdam. The second example focusses on enhancing the habitat suitability of bed protection, while the third example addresses improved connectivity and potential restoration of a brackish wetland adjacent to a recent harbor extension in Africa. The paper illustrates that the design of nature-based infrastructure requires a focus on opportunity creation and restoration of healthy ecosystems. Healthy ecosystems, in turn, help to ensure that benefits accrue to local as well as regional and global stakeholders in a port development process.

1. INTRODUCTION

Building with Nature is an ecosystem-based concept that specifically seeks to use natural materials, interactions and dynamic processes effectively in the design, realisation, operation and maintenance of hydraulic infrastructures (Waterman, 2010; Ecoshape, 2018). As such, it aims to identify and utilize opportunities to benefit nature while undertaking the infrastructural development of ports required to meet the trade, energy and development challenges of the future. This and similar concepts such as Working with Nature (PIANC, 2011) and Engineering with Nature (Bridges et al., 2016) are being applied in infrastructure design and construction around the world. Indeed, the concept is being implemented in practice in the Netherlands through an adaptive "learning by doing" approach. This paper provides examples of the use of nature-based concepts in port design, and highlights the potential of the approach by illustrating the connection of infrastructure design alternatives and the restoration of ecosystem services.

Since Costanza's seminal work (Costanza et al., 1997), we have come to understand that healthy natural ecosystems provide services to individuals and to society (MEA, 2005; Bateman et al., 2013). These ecosystem services are considered to comprise production services (such as food harvesting), regulation services (such as flood control) and cultural services (such as recreation, spiritual significance). The ecosystem services concept is now widely used in identifying, quantifying and evaluating the effects of infrastructure design alternatives on the delivery of services to humans by the affected ecosystem. The use of ecosystem services to assess nature's value constitutes a growing field of research over the past decade, and has been associated with the Building with Nature concept (Smith & Deerenberg, 2012).

[1] Delft University of Technology, Faculty of Civil Engineering & Geosciences, & Faculty of Technology, Policy & Management, Netherlands, j.h.slinger@tudelft.nl
[2] Royal Boskalis, Netherlands
[3] Deltares, Netherlands
[4] Wageningen Marine Research, Netherlands
[5] Institute for Environmental Studies, Netherlands
[6] University of Ghana, Department of Marine and Fisheries Sciences, Ghana

2. EXAMPLES OF NATURE-BASED ALTERNATIVES IN PORT DEVELOPMENT

A first example of a port-related and nature-based infrastructural project is the extension to the Port of Rotterdam, the Maasvlakte II, which is protected from coastal storm surge mostly by sandy dunes. The dune field enhances nature development, recreation and is simultaneously extremely effective and less costly than full protection by breakwaters. However, designing this option instead of a standard breakwater is not available to ports that are located on rocky coasts, or where sand is not readily available or could be damaging to ambient ecosystems such as coral reefs. An example of nature-based methods which could be relevant in such areas, draws on a study of the distribution of cavities (size and connectedness) in the bed-protection at the base of wind turbines. This study by Buijs (2015) investigated which size ratios of the rocks comprising the bed protection were best suited to the creation of appropriate cavities to shelter rock lobster in the North Sea. The outcomes yielded insight in the selection of rock size ratios, and indicated that the habitat suitability for rock lobster could be improved without significant additional cost. In addition to these two more general instances of ecosystem-based infrastructure design, the project "Integrated and Sustainable Port Development in Ghana in an African Context" (NWO 2018), has sought to extend the practice-based learning to a case study in Africa, namely the Port of Tema.

Preliminary investigative work on potential nature-based alternatives in the extension of the Port of Tema was undertaken by Vrolijk (2015). She identified the Sakumono Lagoon to the west of the harbour as a potential site for ecosystem restoration or opportunity creation, and this was confirmed through the stakeholder engagement process undertaken for the Tema case study (Slinger et al., 2018). At present, a multidisciplinary group of students from the Delft University of Technology is currently working with students and professors from the University of Ghana to design a number of interventions aimed at restoring this lagoon, or at the least effectively managing the water quality and habitat loss issues of this wetland area.partners and supporting parties (see Vellinga (2017) for information about the project organization) was held in Delft, the Netherlands, in September 2016 with similar objectives (Vreugdenhil, 2016). The Ghanaian researchers participated in the workshop through an active skype link. The workshop agenda incorporated an introductory presentation on the project, which included the input from the first workshop, a question and answer session, a personal mini-survey over what is (or is not) a sustainable port in Tema or an ideal world. This was then followed by group sessions on sustainable ports and general discussion. Then, the serious game Port of the Future was played as part of the workshop. This was done for parity with the stakeholder workshop in Ghana. Finally participants were asked to reflect on workshop. The outcomes were similar to the first workshop in Tema, namely increased awareness of sustainability and consensus on the most important issues to address via the project. In addition, the connectedness of the network around the project was strengthened, and the inputs and insights from the Ghanaian partners added depth to the discussion on sustainability within an African context.

The Sakumono II Lagoon is located very near to the furthest extension of the new breakwater of the expanded port of Tema. Currently the exchange of water with the sea is strongly curtailed as the connection only occurs through two large culverts extending under the coastal road. Given the anticipated increase in traffic to and from the expanded port it is understood that the road network around Tema will need to be developed. This is already happening with the connection between the Port and the main highway immediately to the north of the city of Tema currently being improved. However, the coastal road cannot lag far behind. Accordingly, preliminary designs for a bridge to replace the culverts and allow the free exchange of marine and freshwater through the mouth of the Sakumono II Lagoon are being investigated in collaboration with scientists from the project Integrated and Sustainable Port Development in Ghana in an African Context. Restoring the connection would restore continuity of water flow and sediment exchange and re-establish a brackish environment. Following Slinger (2000), it is anticipated that after some adjustment time, involving dieback of the freshwater vegetation that has colonized the lower reaches of the lagoon, the flora and fauna characteristic of a brackish wetland would re-establish. However, it is not sufficient to simply restore water exchange. The wetland is subject to high nutrient loading from waste effluent and has experienced decades of sediment infilling owing to reduced flushing and landfill for settlement. Additional interventions such as dredging and nutrient management (cf. Taljaard et al., 2017) would also be required if opportunities for nature, and concomitant ecosystem services benefits, are to be realised. These benefits include increased livelihood fishing opportunities for the fisherman of Sakumono village (to the west of Tema) whose marine fishing grounds have decreased with the expansion of the harbour. Additionally, the residents on the upstream eastern bank of the lagoon may be flooded less frequently. At present they are suffering flooding on an annual basis in the wet season as the constricted mouth causes back-flooding. A nature-based approach focussed on the restoration of ecosystem processes, so as to once again permit the delivery of production and regulatory ecosystem services, would simultaneously restore the aspects of the cultural services. The Sakumono II Lagoon holds spiritual significance for the indigenous peoples of the area.

3. IN CONCLUSION

In this paper, we have illustrated that the design of nature-based infrastructure requires a focus on opportunity creation and restoration of healthy ecosystems. This helps to ensure that benefits accrue to local as well as regional and global stakeholders in a port development process.

References

Bridges, T.S., Banks, C.J., Chasten, M.A. (2016). Engineering with nature. Advancing system resilience and sustainable development. The Military Engineer, January-February 2016: 52 – 54. https://ewn.el.erdc.dren.mil/Publications_files/Publications/Pub%201%20EWN%20TME-JanFeb2016.pdf#view=fit

Buijs, W. (2015). Ecology-based bed protection of offshore wind turbines. MSc thesis, Delft University of Technology, Delft, The Netherlands. http://repository.tudelft.nl/view/ir/uuid:13e98ca5-bc9b-4935-ab85-8aa03593fc88

Bateman, I.J., Harwood, A.R., Mace, G.M., Watson, R.T., Abson, D.J., Andrews, B., Binner, A., Crowe, A., Day, B.H., Dugdale, S., Fezzi, C., Foden, J., Hadley, D., Haines-Young, R., Hulme, M., Kontoleon, A., Lovett, A.A., Munday, P., Pascual, U., Paterson, J., Perino, G., Sen, A., Siriwardena, G., Van Soest, D.P., Termansen, M., (2013). Bringing Ecosystem Services into Economic Decision-Making: Land Use in the United Kingdom. Science 341, 46–50.

Costanza, R., R. d'Arge, R. de Groot, S. Faber, M. Grasso, B. Hannon, K. Limburg, S. Naeem, R.V. O'Neill, J. Paruelo, R.G. Raskin, P. Sutton, and M. v. d. Belt. (1997). The value of the world's ecosystem services and natural capital. Nature 387:253-260.

Ecoshape (2018). Building with Nature, www.ecoshape.nl ; accessed 3 March 2018.

Millennium Ecosystem Assessment, (2005). Ecosystems and Human Well-being: Synthesis. Island Press, Washington, DC.

NWO (2018). Integrated and Sustainable Port Development in Ghana within an African context, Urbanising Deltas of the World Program, URL: http://www.nwo.nl/en/research-and-results/research-projects/i/95/13995.html ; accessed February, 2018

PIANC (The World Association for Waterborne Transport Infrastructure). (2011). PIANC Position Paper 'Working with Nature', www.pianc.org/workingwithnature php ; accessed 19 September 2016.

Slinger, J.H. (2000). Decision support for the conservation and management of estuaries. Final report of the predictive capability sub-project of the co-ordinated research programme. WRC Report No. 577/2/00, Water Research Commission, Pretoria, South Africa.132 pp + appendix. ISBN 1 86845 635 8.

Slinger, J., P. Taneja, B. Kothuis, H. Vreugdenhil, W.P. de Boer, A. Kangeri, M. Koetse, L. Hagendoorn, C. Dorsser, E. Mahu, B. Amisigo, K. Appeaning Addo, T. Vellinga (2018). Designing for stakeholder values in port development in Africa. Presented at PIANC 2018, 7 – 11 May 2018, Panama city, Panama.

Smith, S.R. & Deerenberg, C. (2012). A ranking of coastal EU habitat types using Ecosystem Services. IMARES report C159/12.

Taljaard, S., van Niekerk, L., Slinger, J.H. (2017). A screening model for assessing water quality in small, dynamic estuaries. Ocean and Coastal Management 146: 1-14. http://dx.doi.org/10.1016/j.ocecoaman.2017.05.011

Vrolijk, E. (2015). Ecosystem-based port design. An approach for sustainable development. MSc thesis, Delft University of Technology, Delft, The Netherlands. http://repository.tudelft.nl/view/ir/uuid:5aa8c5bd-37ef-47f1-8fdd-20114ecc576e

Waterman, R.E. (2010) Integrated coastal policy via Building with Nature. PhD thesis, Delft University of Technology, Delft, The Netherlands.

Acknowledgements

The authors wish to acknowledge the Netherlands Organisation for Scientific Research (NWO-WOTRO) for financing this research project (nr. W07.69.206) under the Urbanizing Deltas of the World programme. Sincere gratitude is expressed to the Delft Deltas, Infrastructures & Mobility Initiative (DIMI https://www.tudelft.nl/infrastructures/) Special Project Ghana Volta Delta for their support.

DIFFUSING KNOWLEDGE ON SUSTAINABLE PORT DEVELOPMENT

by

J. Slinger[1], H. Vreugdenhil[1,2], P. Taneja[1], W.P. de Boer[1,2] A. Kangeri[3], B. Kothuis[1], M. Koetse[4], L. Hagedoorn[4], C. van Dorsser[1], E. Mahu[5], B. Amisigo[5], K. Appeaning Addo[5], T. Vellinga[1]

ABSTRACT

This paper adopts theory on pilot projects to devise a strategy for the diffusion of new knowledge and practices from case study research on an African port. Progress in disseminating knowledge and in creating awareness of new practices is reviewed. The efficacy of the strategy is also assessed.

1. INTRODUCTION

Pilot projects are policy instruments for introducing or testing innovative approaches, concepts or technologies, and can be viewed as stepping stones in the development of new practices (Vreugdenhil et al., 2012). Pilot projects can also stimulate the wider acceptance and implementation of new practices. In a study on pilot projects in water management, Vreugdenhil (2010) identified that the diffusion of new knowledge and practices from pilot projects can occur in four ways:

i) Inclusion of the pilot project in routine project practice (routinization)
ii) Replication of the pilot project at other locations, in different social-ecological contexts, but at similar scale (replication)
iii) Scaling-up of the pilot project to larger spatial or temporal scales (expansion)
iv) Institutionalization of the pilot project approach at higher organisational, sectoral or cross-sectoral levels (institutionalization).

Since the main aim of the project "Integrated and Sustainable Port Development in Ghana in an African Context" (NWO 2018) is to develop knowledge and test design practices for integrated and sustainable ports in an African case study, namely the Port of Tema in Ghana, it is feasible to view the case study endeavour as a research pilot project. This allows us to apply the theory on pilot project diffusion to stimulate the wider dissemination and uptake of the new knowledge and practices developed within the research project.

2. DESIGNING AN INFLUENCE STRATEGY

Because of the long term nature and the high level of investment involved with port development, it is unlikely that port development projects will be routinized. Accordingly, we have adopted two of the four mechanisms for diffusing new knowledge and practices as the basis for our influence strategy (Figure 1). We envisage that the port development pilot project to have a potential ripple effect within Ghana itself, resulting in the replication of (elements of) our case study approach within other potential Ghanaian port development projects (Phase 1 to 2 in Figure 1). We do not anticipate an expansion of the pilot project itself as our inclusive, ecosystem-based approach is broad and has already identified and addressed key strategic issues and key stakeholders from the outset (Taneja et al. 2018; Slinger et al., 2018). We actively seek to disseminate knowledge and procedures from the pilot project to institutionalize the approach within African port development (Phase 2 to 3 in Figure 1), by informing and involving national and pan-African decision makers on harbour

[1] Delft University of Technology, Faculty of Civil Engineering & Geosciences, & Faculty of Technology, Policy & Management, Netherlands, j.h.slinger@tudelft.nl
[2] Deltares, Netherlands
[3] Wageningen Marine Research, Netherlands
[4] Institute for Environmental Studies, Netherlands
[5] University of Ghana, Department of Marine and Fisheries Sciences, Ghana

development in our project where possible (indicated by the upper arrows). We expect this form of pilot project diffusion to be crucial to the ultimate impact of the research project, and will support this by specific networking and dissemination activities. In this way other African case studies will be able to add knowledge, concepts and practices in a bottom-up way and through active Pan-African port and shipping networks. Note that whereas Phase 1 and 2 lie entirely within the scope of the project, the full impact in Phase 3 extends beyond the project lifetime.

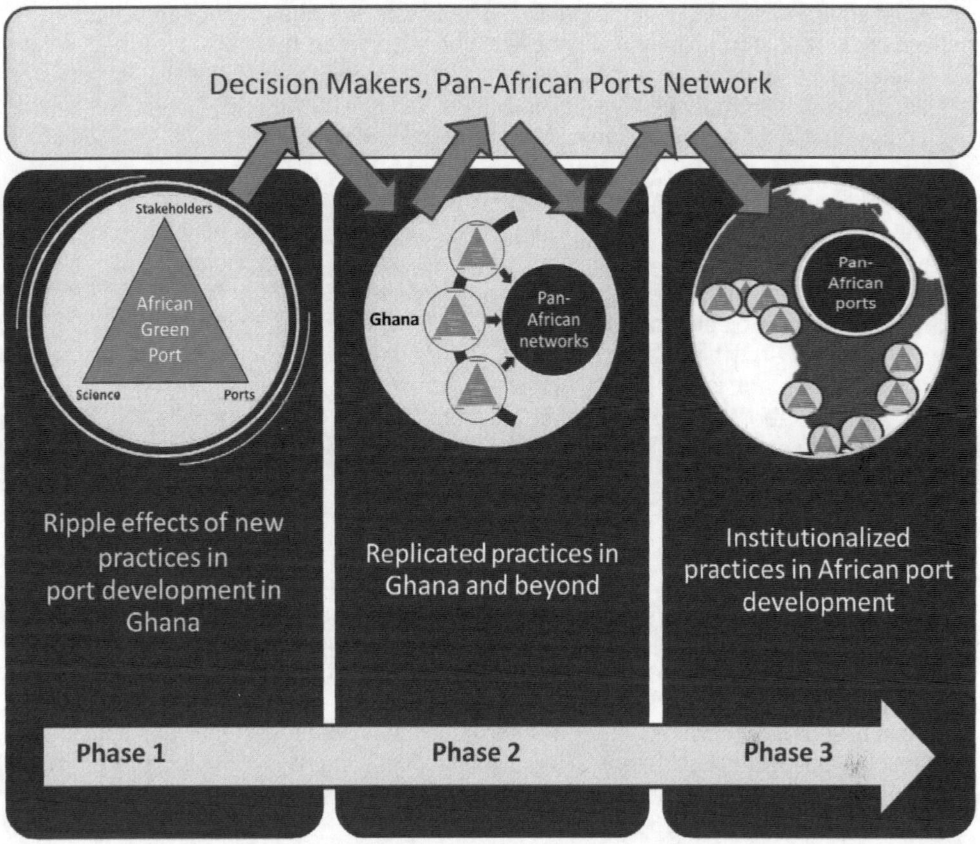

Figure 1: Following the initial development of new port development practices in a pilot project case in Ghana, the replication of practices within Ghana and beyond is envisaged as is the eventual institutionalization through bottom-up adoption and Pan-African networks (NWO, 2018).

The following active communication initiatives were undertaken to enhance knowledge diffusion and the subsequent uptake of research results in the form of new practices for sustainable port design.

- A project website was created http://sustainableportsafrica.com/ with the name “Building a Blueprint for Sustainable Ports in Africa”, for facilitating systematic documentation and widespread and easy sharing of project findings.
- A flyer was printed to promote the project Sustainable Ports in Africa.
- A popular article with the title “De duurzaamste haven van Afrika (the most sustainable port in Africa)” was published online in March 2017 (http://waterviewer.tudelft.nl/world#/de-duurzaamste-haven-van-afrika-1489499617271_) in which various members of the research team (Tiedo Vellinga, Mark Koetse and Arno Kangeri) discussed aspects of stakeholder-inclusive design in Africa.

In addition, presentations were given at conferences attended by port engineers and developers in an effort to sensitize engineers and developers to new practices in port infrastructural design and port development (Vellinga et al., 2017; Slinger et al. 2017; Kangeri, 2017; Vellinga, 2017; Slinger 2017). The conferences included SMTC 2017, 26-28 April, 2017, Singapore, the MARE Conference 2017 in Amsterdam and the African Ports Evolution Conferences for West Africa, in Ghana, and for Southern and Eastern Africa, in Durban, South Africa.

A number of project reports, papers and two masters' theses were also produced within the first year to one and a half years of the project. The Masters theses focussed on aspects not addressed within the main research project, but that had been identified by stakeholders as relevant issues during the stakeholder engagement process. These included the feasibility of an inland water container transport service between Tema and Lake Volta (Toebes, 2017) and the effect of port development on the urban dynamics of Tema city (van den Houten, 2017).

Cross-linkages with on-going projects were also sought, namely:

- Collaboration with the Delft Deltas, Infrastructures & Mobility Initiative (DIMI https://www.tudelft.nl/infrastructures/). The Sustainable Ports in Africa project will contribute to the objectives of DIMI through a Ghana special project on the Volta River Delta and Tema Port in which a multi-faculty education & research collaboration and integrated design approach is applied by a student group.
- Collaboration with researchers from the Deltas, Vulnerability & Climate Change: Migration & Adaptation project (DECCMA http://www.geodata.soton.ac.uk/deccma/).

3. CONCLUDING

On reflection, each of the activities was designed to grow the network aware of the concept of sustainable ports, to share knowledge, disseminate research and initiate uptake activities. This influence strategy has been effective to date, and as a result, the research is already known within a much wider network. The network includes port engineers and developers active in other regions in Africa and incorporates other research endeavours such as DECCMA and DIMI. Indeed, the anticipated uptake and testing is already occurring with the CSIR in South Africa agreeing to comment critically on the applicability of the methods developed within the Tema case study in their port stakeholder-engagement processes.

It remains to influence policy makers towards more sustainable port development and to truly establish a Sustainable Ports in Africa network. This will form the major focus in our ongoing efforts to disseminate new knowledge and practices within the African context..

References

Bridges, NWO (2018). Integrated and Sustainable Port Development in Ghana within an African context, Urbanising Deltas of the World Program, URL: http://www.nwo.nl/en/research-and-results/research-projects/i/95/13995.html ; accessed February, 2018.

Kangeri, A. (2017), Frameworks for Harbor Sustainability in Africa, MARE Conference 2017, People & the Sea IX: Dealing with Maritime Mobilities, 5-7 July, 2017, University of Amsterdam, the Netherlands.

NWO (2018). Integrated and Sustainable Port Development in Ghana within an African context, Urbanising Deltas of the World Program, URL: http://www.nwo.nl/en/research-and-results/research-projects/i/95/13995.html ; accessed February, 2018.

Vellinga, T. (2017), Integrated and Sustainable Port Development: Making a conscious choice for man, environment and economy, African Ports Evolution- west African edition, 5-6 September 2017, Accra, Ghana

Slinger, J. (2017). Sustainable ports in Africa: An integrated "Building with Nature" approach to co-create value at the port-city nexus. African Ports Evolution Conference 2017, Durban, 17-18 October 2017.

Slinger, J., Taneja, P., Vellinga, T., van Dorsser, C. (2017), Stakeholder inclusive design for Sustainable Port Development in Africa, SMTC 2017, 26-28 April, 2017, Singapore.

Slinger, J., P. Taneja, B. Kothuis, H. Vreugdenhil, W.P. de Boer, A. Kangeri, M. Koetse, L. Hagendoorn, C. Dorsser, E. Mahu, B. Amisigo, K. Appeaning Addo, T. Vellinga (2018). Designing for stakeholder values in port development in Africa. Presented at PIANC 2018, 7 – 11 May 2018, Panama City, Panama.

Taneja, P., T. Vellinga, J. Slinger, H. Vreugdenhil, W.P. de Boer, A. Kangeri, B. Kothuis, M. Koetse, L. Hagendoorn, C. Dorsser, E. Mahu, B. Amisigo, K. Appeaning Addo. (2018). Towards a framework for integrated, ecosystem-based port development. Presented at PIANC 2018, 7 – 11 May 2018, Panama City, Panama.

Toebes, K. (2017). Pre-feasibility of an inland water container transport service at lake Volta, Ghana, MSc thesis, Delft University of Technology, Delft, the Netherlands.

van den Houten, J.(2017). A System Dynamics Exploration of Port-City Development: the Case of Tema, Ghana. MSc thesis, Delft University of Technology, Delft, the Netherlands.

Vellinga, T., Slinger, J., Taneja, P., Vreugdenhil, H. (2017), Integrated and Sustainable Port Development in Ghana, SMTC 2017, 26-28 April, 2017, Singapore.

Vreugdenhil, H. S. I. (2010). Pilot projects in water management. Practising change and changing practice. PhD thesis, Delft University of Technology, Delft, the Netherlands.

Vreugdenhil, H., Taljaard, S. and Slinger, J.H. (2012). Pilot projects and their diffusion: a case study of integrated coastal management in South Africa, Int. J. Sustainable Development, Vol. 15, Nos. 1/2, pp.148–172.

Acknowledgements

The authors wish to acknowledge the Netherlands Organisation for Scientific Research (NWO-WOTRO) for financing this research project (nr. W07.69.206) under the Urbanizing Deltas of the World programme. Sincere gratitude is expressed to the Delft Deltas, Infrastructures & Mobility Initiative (DIMI https://www.tudelft.nl/infrastructures/) Special Project Ghana Volta Delta for their support.

AUTHOR INDEX